The Living Origin of
Rocks and Minerals

The
Living Origin
of Rocks *and*
Minerals

Walther Cloos

Floris Books

Translated by Katherine Castelliz and
Barbara Saunders-Davies

First published in German as
Lebensstufen der Erde. Translated from the
second German edition published in 1970
by Verlag Freies Geistesleben GmbH, Stuttgart

First published in English in 1977
by Lanthorn Press as
The Living Earth: The Organic Origin of Rocks and Minerals
Second edition published in 2015 by Floris Books, Edinburgh
Second printing 2023

British Library CIP Data available
ISBN 978-178250-173-2

Contents

Publisher's note

Walther Cloos wrote this book in the 1950s. As a geologist and mineralogist he made the bold and pioneering attempt to reconcile Rudolf Steiner's spiritual insights into the evolution of the Earth with geological phenomena. Since this book was written geology has moved on, as has some of its terminology. Nevertheless, we hope that this classic text provides interest, historical background and stimulates further research.

Preface

In this book we have attempted to look at scientific facts concerning the world of rocks in the light which Rudolf Steiner has shed on all spheres of human life and endeavour in his anthroposophy. Seen from this point of view the Earth is a living organism whose life is interwoven spiritually with the life of the heavenly bodies in the solar system and the whole realm of the stars. What has developed gradually within this life as the kingdoms of nature – mineral, plant and animal – are stages left behind by the being which is the culminating goal of all Earth's existence – the human being.

The author is aware that his attempt to arrive at such a picture of the world of rocks contradicts the view commonly held. The fact that this view has lead to a rape of the Earth and the threat of atomic death may excuse the attempt with all its shortcomings. If we would prevent the premature end of the Earth through arbitrary and thoughtless actions of the present day, we must come to a new conception of the human being and of the Earth.

I am greatly indebted to the work of Guenther Wachsmuth especially his book *Die Entwicklung der Erde,* which deals with the time element of our subject matter. I would like to thank all those friends and colleagues who, through discussions over many years, have contributed consciously or unconsciously to bringing this book to birth.

Walther Cloos

Introduction

The wayfarer who turns his gaze from the abundant life of plants and animals to the solid mineral earth, may not always realise that traces and remains of bygone life are to be found in the rocks and screes. Only in comparatively few places on the surface of the Earth can one see that the rock contains fossilised remains of animals resembling mussels and snails. These and related forms constitute by far the greatest part of what is clearly recognisable as traces of life. Furthermore it is evident that the rocks containing such traces of life are mostly limestones, calcareous shales, marls, or shales interbedded in limestones.

If one takes the trouble to examine limestone under a magnifying glass or a microscope, it is easy to see that many consist entirely of the minute tests of protozoa (Protista), organisms which inhabit our oceans to this day. The zoologist Richard Hertwig writes: 'No group of animals has contributed so much to the formation of new strata in the present and in the past as they have.'

We find the protozoa in layers between the coal seams as fusulinid limestone and we find them above all in the Muschelkalk, in the Jurassic limestones, the Chalk and the nummulitic limestones of the early Tertiary period.

If we think of the plateau mountains of the Jura, the white chalk cliffs of the south coast of England, the Baltic coast near Rügen, or the endless central plains of the American continent which are underlaid by thousands of metres of chalk, we have to realise that these tremendous layers of limestone are largely made up of myriads of minute organisms that have perished.

Certainly, we still find today in the oceans, legions of these little creatures which deposit their calcareous shells (tests) to form the ooze of the deep sea bed, but we can no longer observe the formation of tremendous calcium deposits like those of the past. It is obvious that the abundance of life is reduced, that the intensity of these processes has slackened off and also that these little creatures tend to be much smaller than they were in the past.

All these observations suggest that, at the time of the formation of these limestones, life processes were far more all-embracing, that the oceans, where they occurred, predominated and that the Earth as a whole was in a more watery state. This idea is substantiated by the fact that in some of these limestone strata we find the remains of large animals which at first somewhat resemble fish, newts, crocodiles, and later lead on to the strange forms of the giant saurians.

From the change in these forms of animal life we can see that this time was a significant transition period. The dominion of the sea was broken by the gradual formation of solid land. As we have seen, it was the tiniest organisms which were mainly responsible – at least in these limestone strata.

If we turn from the Chalk, Jurassic limestones and Muschelkalk, and go deeper into the Earth, we observe that the limestone layers become scarcer and we find rocks of which the blue roofing slates are a typical example. In these slates we find remains of marine animals, many of which today are extinct. From these animal forms it is clear that, though water played an important role when the rocks were formed, it must have been a very different sort of ocean, for it not only left limestone beds but also clayey mud, for slate consists mainly of aluminium silicate with the admixture of carbon and iron. The slates alternate with sandstones, or sometimes with limestones or calcareous shales. As one penetrates deeper, approaching the earliest times of the Earth, their place is taken by the glittering mica schists and gneiss. At the end of the schist/slate period something very special appears – the

Coal Measures. They indicate a more vegetative life in contrast to the animal life of the limestone period.

If one goes still deeper and further back beyond the schist and gneiss periods, one comes to the real primal rock exemplified in granite and its related rocks. These rocks are granular and more or less uniformly composed of quartz, mica, hornblende and feldspar. It is important to bear in mind that the oldest rocks like granite have a granular structure, the schists and slates and shales are foliated and limestone is characterised by thick beds.

In the deeper layers of shale, embedded in between limestones, sandstones and salt deposits lies one of the most important remains of ancient life – oil. This however constitutes only a part of the oil. Oil reappears in the Tertiary epoch. It is the same with coal; it reappears in the Tertiary as lignite or brown coal.

In the depths where the rocks begin to become crystalline (mica schists and gneiss, Plate 1 – see inside cover) the last traces of life in the form of fossils and impressions disappear. But even here we find between the strata huge beds of marble or dolomite (a double carbonate of calcium and magnesium) which point towards tremendous plant and animal life processes and may have been produced by something similar to our present-day algae.

With this rough sketch of the three principal periods of the Earth on which our further studies will be based, we have by no means covered all the rocks that constitute the hard crust of our Earth. To these must be added the group known as igneous.

These igneous rocks of which we will only mention granite, the greenstones, porphyry, trachyte, plateau basalt and lava (Plates 2, 13, 14) have from the most ancient times domed up, broken through and overflowed the layers of sandstones, slates and limestones. They are closely associated with the formation of mountain ranges, though of course not all mountains are of volcanic origin.

To get an idea of the mass of the various strata we need a few figures. The 'Father of Geochemistry', F.W. Clarke, determined this for the known part of the Earth's crust to a depth of 16 000 metres (53 000 ft). This depth can be considered as known since oil borings have reached 6 000 metres (20 000 ft) and more in many parts of the world, and tremendous folding has brought the deep rocks to the surface in mountain ranges many thousands of metres above sea level. If one adds the sea and the atmosphere to these rocks, one finds that the comparative weights of the Earth's crust are: rocks 93.06%, water 6.91% and atmosphere 0.03%.

The rocks forming this 93.06% are about 88.4% igneous rocks, 3.7% schists, slates and shales, 0.7% sandstones and only 0.2% calcareous rocks. It is evident that the rocks we have classified as of granular structure, far outweigh everything else.

The relationship between the igneous rocks and the sedimentary rocks (sandstones, limestones, slates, etc.) appears to be very different if one only looks at the surface of the Earth, as some three quarters of it is covered with sedimentary rocks and only one quarter has the igneous rocks exposed.

Again if we consider the thickness of the individual strata their massiveness is such that one simply asks: Where has it all come from? If one looks at all these phenomena with an open mind one is struck with amazement and wonder, and it is this attitude of mind which should be the starting point for true scientific enquiry.

The question of the origin of the rocks is as old as science itself. The age old dispute between the neptunists who maintained that everything arose from water, and the plutonists who referred everything back to primal fiery and molten processes, is not settled even today. The modern assumption is basically that rock materials originated from a condition of incandescent gas. The gas would then have cooled to a molten mass producing a primal magma – somewhat resembling basalt. After further cooling, water would have formed on the surface crust, and from its interaction with the boiling magma of the interior, the enormous variety of rocks arose.

This theory considers the sedimentary rocks of the slate/ limestone times as simply made up of debris and dissolved matter ultimately derived from the cooled magma. Even the enormous masses of calcium used by the Protista to build their tests are explained as originating from the calcium content of the primal magma dissolved in the sea water.

This theory of the origin of rocks leaves a cardinal question unanswered: *Where did life come from?* The origin of life is a question to which modern science has absolutely no satisfactory answer. There was, and still is, a theory about bacteria in the universe which, through pressure of sunlight, entered the cooled atmosphere and so formed the first life on Earth. More modern theories postulate a spontaneous generation of simple proteins from methane, carbon dioxide, ammonia and water in the warm atmosphere through electrical discharge. From these bacteria-like or virus-like primal proteins began a development which, assisted by influences from outside, finally produced man.

There are several theories of spontaneous generation but they all forget the most important thing. Not only is there life on Earth, but there is ensouled life, and furthermore ensouled life that is aware of itself – the human being.

If one considers human beings on earth endowed with soul and spirit, and tries to reconcile this with scientific views of geology, one comes to an impasse. In practice this is because today we normally only observe physical and chemical processes at work in the rocks, and no longer life processes – albeit the calcareous mountains bear witness to former life processes.

In plant, animal and humans the mineral is a living, feeling and thought-bearing substance and as such it is not only subject to the chemical and physical laws that hold sway in the world of rocks. But when a plant, animal or human being dies, then the chemical and physical laws take over and dissolve the body.

If there were a tiny creature able to study the decaying processes of a corpse, yet unable to see the whole and understand the organic origin of this corpse, it would be in the same situation as present-day man *vis-à-vis* the rocks of the Earth. In the same way as this little creature would only find physical and chemical laws active in the decaying corpse, so in the world of the rocks today we can only observe physical and chemical processes. The question of the origin of the mineral mass would preoccupy it when it came across the quantities of calcium phosphate which constitute our skeleton. It would also find in the corpse iron, magnesium, sodium and carbon – all materials which are widely distributed in our rocks. If, however, this little creature observed that all these substances were arranged in a certain *order*, that there were layers and zones in which one or another substance predominated or receded, it might arrive at the conclusion that behind this order there was an 'idea', a principle which had ordered it.

Facing the world of rocks we find an ordering. Goethe in his geological studies came to the conclusion that the perfect classification is not made by man; nature herself has laid it out before his eyes. This classification or order – obviously recognised by modern geology – has to be further pursued without letting our gaze be dimmed by preconceived ideas and trying to explain the past from the present.

Rudolf Steiner in his cosmology presented the 'idea' behind the formation of the world of rock, so that one can see how the mineral deposits of the Earth are the transformed remains of former life.

This former life, which had no resemblance to that of today, has transformed itself into the present higher kingdoms of nature but it has also left its 'corpse' behind. If we would not just dissect this corpse but look at it as a whole, then the original magnitude of this ancient life would emerge and the layers of the past would become the leaves of a book which we could learn slowly but surely to read.

1. An Outline of the World of Rocks

Wherever the oldest crystalline rocks of the Earth are found, they show very little variation the world over. Be they on the surface or deep within the Earth, they are simple combinations of quartz, feldspar and mica or hornblende. The differences are mainly in the size of the crystalline granules and the colour. The colour depends on the relative amount of mica and feldspar, and on certain metals, mostly iron and sometimes manganese.

This uniformity is very characteristic of these oldest times. Differentiation only starts later. Thus we have a time in the history of the Earth when no slate, limestone or coal were formed. There was only one basic process which we can call *granite formation*.

Immediately above the granite and also inter penetrating and surrounding it one finds the first modification of the crystalline structure in gneiss. Gneiss is not so uniformly structured as granite. Though still granular, it is layered. It can be relatively homogenous with layers caused by the little flakes of mica, or it can consist of layers of quartz, mica and feldspar more or less separated from each other. One has the distinct impression that what has developed as mica has given the whole rock its structure. This 'mica-effect' is the hallmark of the development of the following period just as the granular structure was for granite.

A further development of this separation is to be found in mica schist where the feldspar has almost completely disappeared and the quartz lies in thick layers in the schists. At the same time a further development of granite itself (quartz, feldspar and mica or hornblende) takes place. These new granitic rocks, the greenstones – syenite, diorite, diabase, gabbro and serpentine – contain no quartz. Syenite and diorite are still very granitic so that the layman can easily confuse them. All these rocks consist of feldspar and mica or hornblende, and can occur between gneiss and mica-schists. Magnesium plays an important role. The dark iron-containing mica of syenite, and the green hornblendes and related minerals are primarily magnesium ferro-silicates. These latter give the rocks, especially diorite, diabase and gabbro, their green or grey-green colour.

There appears to have been a conflict at this stage between two formative tendencies – one to form laminated schistose rocks, the other granitic granulated ones. So it happens that syenite, diorite, diabase and gabbro can appear laminated or granular.

Thus it is incorrect to speak of granite as an igneous rock as used to be done, for what we call today the 'igneous process' has itself gone through a *development* like everything else in nature. The beginning of this development is to be found in gneiss, in the crystalline schists and the greenstones. Later we shall see that it is from these rocks, especially the greenstones, that the true igneous rocks such as the basalts and other lavas develop. As the third stage of this development, which one might consider as a differentiation of tendencies within the primal granite, we have the appearance of sandstones, slates and limestones. Granite is composed of quartz, feldspar, mica and/or hornblende. What was quartz now is sandstone, what was mica or hornblende appears as slates and shales, and in the place of feldspar we have limestone. (Mica contains much aluminium and magnesium; and in granite, feldspar has something of the character of limestone). The threefold composition of granite becomes the great threefold division of sedimentary rocks – sandstones, shales and limestones.

Granite appears as a sort of *prima materia,* a forerunner containing the possibilities of future development. The processes of rock formation become more and more diversified and as the primeval unity disappears, the primal granular rocks also go through a further development. We have outlined how, as the quartz more or less disappeared, granite became syenite, diorite and all the other granular greenstones.

After this greenstone stage quite new forms of rock appear – the porphyries (Plate 2). These porphyries can sometimes look like granite but are generally more finely granulated or even glassy, with the feldspar in large crystals. They contain little or no mica, but hornblende and quartz instead, which in true porphyries are blended in a glassy mass.

These porphyries appeared about the same time as the first sandstones, shales and limestones. Later on limestones take precedence over sandstones and shales. We know that limestone is the result of animal life. It is thus a true sedimentary rock, as are also the accompanying sandstones and shales. At this stage the 'eruptive rocks' trachyte, basalt and lava also occur. Their origin will be discussed later.

We have endeavoured to reduce to its essence the great complexity of rocks. All rocks are distributed over the Earth in distinctive ways and there are large areas where one or other are entirely absent. Thus it is clear that the original uniform process that produced granite became differentiated, so that in one area the development was in the direction of laminated rocks (schists, etc.), in another towards limestones and in another towards the greenstone-type rocks.

We may tabulate this as follows:

1. *Granite:* granular.
2. *Gneiss:* beginning of laminated structures;
   greenstones: both granular and laminated.
3. *Sandstones, shales and limestones:* the granite formation has broken up. Slates/shales predominate.
   Porphyries.

4. *Sandstones, shales and limestones.*
   Limestone predominates.
   Trachyte, basalt and *lava.*

Based on the composition of granite we have:

quartz	mica	feldspar
	hornblende	

In the sedimentary rock development this becomes:

sandstones	shales	limestone

The corresponding 'igneous' rocks according to the above presentation are:

(Older formations)
granites	greenstones	porphyries

(Younger formations)
trachyte	basalt	other lavas

Such an arrangement does not correspond with that of orthodox geology. It does not attempt to be complete, but it deals with the basic phenomena. Such an outline can be a great help in finding our way later through the complexity of the phenomena. If we only consider these simple phenomena, of course we come up against a difficulty when we considers limestone and coal formations. We know that limestone is the product of animal life and that coal originates from plant life. Both deposits are to be found in old and young strata and always lie between apparently entirely inorganic layers. These inorganic layers of sandstone, shales or igneous rock appear as though they originated in the same way as what we observe today when sand or mud are deposited in water, or the molten lava flows from a volcanic crater. But today sand no longer makes sandstone and clay does not form shales, neither does lava turn into granular crystalline rock.

We have already seen that the enormous quantities of limestone can only be understood if we postulate a far more intense life than exists today in our seas.

Likewise with coal. It has been calculated that a tropical rain forest of today would, if suddenly destroyed by a natural catastrophy, only yield a layer of coal a few millimetres thick. Here we are faced with the question: What an extraordinary plant life it must have been to yield hundreds of metres of coal.

Today we know of no life process of such intensity that it could form hundreds of metres of limestone or coal within any conceivable time. Were there not perhaps quite different conditions then on the Earth which do not follow the natural processes of today? Modern research does not offer any satisfactory answers. The answers given speak only of immense periods of time, high pressure in the interior of the Earth, high temperatures which have transformed materials. Experiments appear to support these theories. But experiments alone do not satisfy, since they take as their point of departure lifeless substances which are taken to be the origin of everything.

Nowhere can we observe that life develops from inorganic substances. On the contrary we see that life decomposes *into* inorganic substances which do not come back to life. Living organisms can take up inorganic substances but not all these substances need necessarily have been of mineral origin.

In modern biology the experiments of Otto Heinrich Warburg show that plants have the capacity to collect extremely finely dispersed substances in the atmosphere or soil (physicists speak of highly ionized conditions) and condense them into measurable quantities. This has been made use of in biodynamic agriculture which shows quite clearly that the living soil with its plants has the capacity to replenish its basic nutrients without the application of mineral fertilisers. If soils treated biodynamically for forty years are tested, phosphorus, potassium and nitrogen are found to be present in the necessary quantities in both soil and plant. Although the plants appear to take up these substances year after year from the soil, yet they do not diminish.

The trouble is that through the use of mineral fertilisers the plant can get out of tune with the original order of life. A plant which, for decades and many generations, has been bred on exhausted mineral fertilised soil has simply lost the capacity to collect its own substances; it has degenerated, and follows those laws which the experimenter has imposed upon it. This is the secret of one of the fundamental errors of our time, and this error casts its shadow even into our ideas of the past evolution of our Earth.

If these ideas lead to the insight that *first* there was life and from it condensed the inorganic and lifeless, then we are in a position to take the researches of Rudolf Steiner on the past of the Earth as the foundation of our further observations.

2. Previous Stages of the Living Earth

If one presumes – according to Rudolf Steiner's spiritual investigation – that first there was life, and the dead mineral matter was cast off as part of a dying process, the question arises: From where does life come? The answer to this question does not come within the scope of our discussion for we are dealing primarily with the origin of the mineral kingdom and its forms. For the answer to this question we would refer to Rudolf Steiner's cosmology in his book *An Outline of Esoteric Science*. The past stages of the Earth which Rudolf Steiner described there must, however, concern us, for they are the basis for those vital earth-processes that lead the not-yet-tangible into the mineral, visible form. These processes also underlie Ernst Haeckel's biogenetic basic law which states that the development of individual organisms repeats the development of the world as a whole. (As formulated by Haeckel the development of the embryo is the repetition of the development of the whole tree of life.)

The prehistory of the Earth is repeated in the embryonic development of plant, animal and man; only then can they appear in their present form. This recapitulation is obscure and not directly observable, for it takes place in the smallest and most delicate realms of life. Rudolf Steiner suggested that through a study of the physiological processes of embryonic development it would be possible to gain a picture of the whole past development of the Earth. It is a task for the future to discover the secrets of this past Earth development especially through the study of human physiology.

With regard to the mineral kingdom it is different. Since its forms are really cast-off remains of life we are dealing with something that is finished and not with something in the *process of becoming*, the physiological processes of which we could investigate. We may however be sure that in this finished work which has come to a certain state of rest as mineral kingdom we will be able to find images which can tell us something of the life processes of the Earth's past as described by Rudolf Steiner.

What we find in the mineral kingdom shows that in the younger layers of sedimentary rocks remains of animal life-processes predominate, while in the older ones we have mainly remains of plant life-processes (coal). In the oldest rocks all traces of life are completely absent.

This only informs us that some of the living organisms were sufficiently dense to remain preserved. It does not mean that no other forms of life were present. The apparent absence of traces of life in the primitive rocks is only an absence of those forms which we know as the manifestation of life in later times. How can one relate the life processes of the earliest beginnings of the Earth – before mineralisation – to what later is precipitated and becomes rock? We can gain some idea by observing how in animals and humans there is the metabolic *processes* on the one hand, and on the other, the *end product* of these processes deposited as substances such as bones, nerves, brain, etc. These are, by comparison with organs such as the liver, lungs, kidneys or heart, endowed with only minimal life. For example, nerves and parts of the brain damaged through injury are not capable of regeneration.

In the plant world we have comparable processes in the flower and seed formation on the one hand, and wood and root formation on the other. Flower and seed formation correspond to metabolism in man; wood and root to bones, nerves and brain.

From such living examples we can see how life processes are related to structures which have been cast off from the living. In other words, one cannot adequately consider the mineral kingdom without taking into account the plant, animal and human kingdoms. These have, in the course of their development, deposited the mineral kingdom in the same way as human beings and animal form their bones, nerves and brain, and as trees form their wooden trunks.

Anyone who considers the mineral kingdom alone without including the life processes which have formed it, is like someone trying to explain bones, nerves and brain in isolation, or like a timber expert who does not ask how or where timber grows.

In 1923 in a notebook Rudolf Steiner wrote something which links the mineral world with the development of the kingdoms of nature and man. What the note contains is the result purely of spiritual investigation and expresses in a few words the development of the whole of nature and humankind. Rudolf Steiner referred to this theme in numerous lectures in the preceeding decades, but nowhere is it dealt with thoroughly and in detail. One has the impression these notes are a sketch for a cycle of lectures which were never given. The entry reads as follows:

In porphyry the world-plant-animal expires,
then in schist the plant-nature,
and in limestone the animal-nature.
In salt the universal human being is extinguished.
The opposite pole is sulphur in which the mineral is burnt and consumed.

Mineral is consumed in sulphur,
Plant spreads warmth in layers in schist,
Plant-animal in-grains feeling in porphyry,
Animal conserves forms in limestone,
Man condenses thoughts in salt.

With fiery sulphur man enters the earth,
In schistose layering he makes himself at home on earth,
By awakening awareness (porphyry-like) he finds his limbs and
forms himself as man in limestone,
To found the ground for thought in salt-depositing. ★

Here a concrete connection is established between life
processes and rock formation. New concepts emerge such as
'plant-animal' and burning up of minerals in 'sulphur'; and then
there is the significant pointer as to how man is connected with
the five main processes active in the formation of rocks. We shall
explore further these new concepts later in order to understand
what is at first sight new and rather startling.

The first part of the above text speaks of a dying away as the
basis of the formation of rocks. One should not imagine that
these 'world-plant-animals', 'animal-beings', 'plant-beings' and
'human-beings' were single living individuals which died in
masses and formed rock out of their bodies. In this sense the
world-plant-animal and the being of plant, animal and man, is
referring to something universal which contains these kingdoms
of nature but as yet is not differentiated individually. This 'being'

★ Notebook Archive No 310 of 1923; first published in 1927 in the weekly
  Anthroposophie. Wochenschrift für freies Geistesleben, No. 41.

Im Porphyr erstirbt das Welten-Pflanzen-Tier –
dann erstirbt im Schiefer das Pflanzenwesen
im Kalk das Tierwesen –
und im Salz erlischt das Menschenwesen –
Der andere Pol ist der Schwefel in dem das Mineral verbrennt.

Mineral *verbrennt im Schwefel*
Planze *schichtet Wärme in Schiefer*
Pflanzen-Tier *körnt Empfindung im Porphyr.*
Tier *bewahrt Formgebilde im Kalk*
Mensch *dichtet Gedanken in das Salz.*

Sulphurisch tritt der Mensch ins Irdische
Schiefrig macht er sich die Anpassung ans Irdische zurecht
Empfindung weckend (porphyrartig) gliedert er sich und formt sich menschlich im Kalk,
um in Salz-Einlagerung die Gedankengrundlagen zu schaffen.

is all-embracing in the sense that its life forms the whole Earth. The Earth itself was once living as a world-plant-animal, it was entirely plant being, animal being, human being. As this original all-embracing entity dies and disappears, there emerges that which later becomes rock and the *individual* forms of plants, animals and man.

In order to imagine spatially this primeval condition of the Earth, one has to picture the Earth at that time as having a circumference extending far into the planetary spheres. In a lecture to the workmen at the Goetheanum Rudolf Steiner stated that the Earth was enormous and 'in fact approximately as large as Jupiter'.* The Earth has shrunk tremendously, it has dried up, withered and its former life is dead. Originally this life was active in the periphery. In the process of dying and shrinking, the residues of this life rained down from this periphery and later condensed and hardened as rock.

The beginning of the shrinking of the Earth goes back to a far distant past. The major stages of the Earth evolution described by Rudolf Steiner in his *Outline of Esoteric Science* as the Saturn, Sun and Moon conditions were extended spacially to the present-day orbits of Saturn, Jupiter and Mars. In this process of contraction our present planets were separated off and left behind, as was the Sun. The original body of warmth of Old Saturn contracted and became the gaseous body of the Old Sun and this contracted further to the fluid body of Old Moon. Before each new development, a recapitulation of previous stages of these past incarnations of the Earth took place according to the biogenetic law.

At the beginning of our present Earth development there was also a recapitulation of these previous stages. Then that which was peculiar to our Earth emerged, and this led finally to the formation of solid rock. We are dealing with rhythmic processes or repetitions each on a higher level than the last.

The first beginning of the Earth is a pure heat phenomenon.

* *From Crystals to Crocodiles*, lecture of Sep 23, 1922.

Warmth-substance, sacrificed by a hierarchy of creative beings, is the origin of the material development of the Old Saturn stage. The 'result' of this Saturn-development was the emergence of the germinal stage of human beings in the form of entities without individual life. This life was only incorporated on the Old Sun. At the same time heat densified and rarified. Gas and light emerged. On the subsequent Old Moon stage human beings were endowed with awareness, the gaseous was densified to the fluid state, and in the rarefying process, sound came into being. Only during the Earth stage did human beings begin to develop individual consciousness, and the *solid* separated from the fluid. Further there emerged the form-principle, born of sound, capable of shaping the solid physical substance.

So there was a sequence of evolutionary events whose nature and complexity unfolded rhythmically, rising higher at each repetition. The original heat substance which really was the beginning of what later became the solid mineral, experienced during the second stage an infusing of life. The plant kingdom was conceived. During the third stage this substance was endowed with awareness and became the first beginning of the animal kingdom. Only during the fourth stage, on the Earth, did this substance become bearer of conscious thought, and the human being arose.

This is a very short outline of the development of our Earth in pre-geological times. The subject needs careful and detailed study before it can be truly assessed.

When Rudolf Steiner in his note wrote of the expiring and extinguishing of the collective beings of plant, animal and man he meant that from the all-pervading, living plant substance the individual plants began to emerge. The same applies to the undifferentiated living and feeling animal substance from which individual animals began to arise.

In the case of man, Rudolf Steiner did not use the word expire or die away, but extinguished. This suggests that what has been extinguished can also be rekindled. For it is the task of human beings to bring down to the Earth in full waking consciousness the essential reality in which they live before birth and after death.

This passage in Rudolf Steiner's note recalls a saying from the fragments of Novalis: 'When a spirit dies, a man is born: when a man dies, a spirit is born'.

Man is 'a citizen of the universe and a hermit on earth' as Rudolf Steiner once said. Human beings are really in their true element before birth and after death. With birth the human being who was formerly embedded in the whole universe is extinguished. But on the Earth humans can try to regain in their consciousness what they experienced before birth and will experience after death. They can rekindle their true being. Originally human beings were permanently embedded in the life of the whole universe. When the phenomenon of birth began – as humans emerged from the floating existence similar to what they experience today in the womb before birth – they appeared in solid physical form. At the same time the rocks and salts solidified out of the fluid Earth.

So we see there were conditions on Earth when certain groups of substances were really imbued with life. Substance was fluid, permeated with warmth, air and light. This was the primal albuminous atmosphere of which Rudolf Steiner spoke in his lectures *Mystery Knowledge and Mystery Centres*. The substances which later hardened into rock were in solution in this primal albumen. What was precipitated as mineral substance from the life of the world-plant-animal in its dying away became *porphyry*. What was precipitated from the all-embracing plant nature of the Earth became *laminated rock* (schists, slates, etc.). What was precipitated from the animal nature became *limestone*. Only after all these mineral substances were precipitated from the albumen and their true being had died, did the present-day forms of life appear in all their diversity.

Finally, with the appearance of man the *salts* were precipitated out of the 'universal fluid', and at the same time the still soft rocks began to harden. A surviving remnant of this universal fluid, this original albuminous atmosphere, is the ocean with its high salt content. Another is the amniotic fluid in which humans and animals float before birth – this fluid which contains salt, like seawater, as well as a high amount of sugar and albumen.

Animals and humans have taken into themselves in their blood something of this primal albumen. The salt content of the blood resembles that of the sea. Here the higher organisms have retained something of the original living and sensitive substance from a past stage of the Earth development in order to transform it into the foundation of their awareness and conscious life.

In the last sentence of the first part of the above-mentioned notebook entry Rudolf Steiner wrote of 'sulphur in which the mineral is burnt and consumed.' This terminology is foreign to modern abstract thinking. His concept of sulphur has little to do with the substance sulphur we know as a mineral. 'Sulphur' is an old alchemical term signifying fire or heat. This idea of sulphur referred to a process which took place in heat. One saw a living 'sulphur' for instance in the flowering of plants. Rudolf Steiner used the term to indicate the living fire in which the minerals are consumed.

This implies that through a living heat process certain mineral substances – for instance, silica, aluminium, magnesium, etc. – began to separate out of the primal albumen as definite substances with firm outlines. This does not mean however that crystals appeared.

In this burning process the first beginning of our later minerals appeared, especially those found in granite and the greenstones. In other connections Rudolf Steiner spoke of this burning-up process as an all-embracing flowering process of the whole Earth. He spoke of a world of *mineral-plants* as a repetition of a previous stage of development of the Earth.

Later, as we go into detail, we shall see that the flower and plant nature of this 'burning-up' is still to be discovered pictorially in some rocks.

This mineral-plant-world is very ancient and existed already on the Old Moon stage of the Earth (see *An Outline of Esoteric Science*). Then it was still living. Now, in the repetition on our Earth, it dies and in so doing forms granite and greenstones, the firm foundations of our Earth.

3. The Character of the World of the Mineral-Plants

In the last chapter we spoke of a past stage of life on our Earth which came to its fullest expression during an earlier incarnation of the Earth which Rudolf Steiner in his *Outline of Esoteric Science* called the Old Moon.

This Old Moon was a 'fluid body' which did not yet contain any hard minerals but was through-and-through rampant, living growth. Rudolf Steiner described the mineral-plant world of the Old Moon as an inwardly proliferating, peat-bog like substance. This of course only gives some sort of an idea as our modern peat-bogs are not 'inwardly proliferating'; their only life is on the surface and the edges. These modern peat-bogs are an earthly phenomenon which only appeared later when the real life of the fluid Earth was confined to the surface and the dead substances had sunk to the bottom and later hardened into rock.

Another modern phenomenon that is perhaps nearer to the rampant growth of the Old Moon is the way a cancerous tumour grows. Here, in a pathological way, we have the remarkable fact that cell growth is no longer controlled by the whole organism with its differentiated structure, but becomes self-willed and runs amok. What is significant is that a cancerous tumour often shows embryonic forms which belong to an earlier stage of development of the organism. In this disease therefore human beings, in a way, slip back into an earlier stage of development, something of the Old Moon nature appears in the body and humans are no longer able to control growth into definite organs and tissues.

The rampant growth of the mineral-plant world of the Old Moon showed a distinct differentiation which came about through the fact that this Old Moon at the beginning of its development included what today circles the Earth as the Sun. It was a heavenly body that consisted of today's Sun, Moon and Earth. In this state the Old Moon *recapitulated*, to begin with, those earlier conditions of the Earth that we have called Old Saturn and Old Sun. During the recapitulation of the Old Saturn condition the life of the mineral-plant was permeated with warmth – was *flower-like*. Thick cushions of tiny blossoms emerging from the dry rocks in high mountains, surrounded by flying insects, give us a picture of the life processes which Rudolf Steiner described. If we can picture myriads of such flower cushions interpenetrating each other luxuriantly, it gives us some idea of this remarkable abundance of life.

This flower quality of the life of the Old Moon which was a repetition of the Old Saturn development gradually became more treelike. It did not, however, form branches and leaves but life-forms resembling the annual rings of our present-day trees. Wood and rootlike forms appeared which densified to a hornlike consistency and then dissolved and flowed away again. These fibrous and layered forms of life were not hard or clearly defined but were in a state of continual change or flux. This treelike vegetative growth was a recapitulation of the Old Sun development of the Earth.

With these two recapitulations we are dealing only with mineral and plant processes and forms; there is nothing of an animal nature present yet. Then comes the separation of Sun and Moon: the Sun becomes an independent heavenly body around which the Old Moon revolves. This Old Moon now begins its own proper development and a part of the substance of the mineral-plant becomes endowed with awareness. The new kingdom of the plant-animal appears. But here too, life is an ever-flowing process, never taking on fixed forms. The Sun, acting on the Old Moon from outside, effects a sort of fertilisation in the life of the animal-plants.

In the whole of the life of the Old Moon there are no fixed permanent forms, only densifications to a horny consistency which soon dissolve and return to a state of flux. Towards the end of the Old Moon development, of which we have given a brief account in order to make the later stages of Earth development comprehensible, the two heavenly bodies of Sun and Old Moon reunite and pass away into a state of rest, a *pralaya*.

In the beginning of the Earth development all the previous stages of Old Saturn, Old Sun and Old Moon were recapitulated like with the biogenetic law, before the Earth development proper began. The Earth, which initially was a single heavenly body containing Sun and Moon, first went through a condition of pure warmth which was characteristic of Old Saturn, progressed to the air-light body of Old Sun, and finally to the fluid body of the Old Moon. These three stages were also separated by rest periods. After the third rest period there began the fourth part of the Earth development which had as its goal the densification, step by step, into three dimensional, physical, material existence of all that had gone before. Within this fourth stage of Earth evolution, the Old Saturn, Old Sun and Old Moon conditions were repeated with enhancement and metamorphosis. These repetitions were also separated by rest periods. After the third rest period the true physical development set in and led to the solid crystalline state of the present time after again recapitulating the Old Saturn heat condition, the light-air condition of the Old Sun and the fluid Old Moon condition.

Rudolf Steiner described this threefold repetition in his book *Cosmic Memory* in considerable detail. We have outlined the essentials in order to give some idea of the extraordinary complexity of the development which preceded the physical material emergence of the mineral kingdom.

In this threefold recapitulation on the Earth, the world of

the mineral-plants reappeared passing through all stages of densification from pure warmth, through the gaseous and fluid states, to the fourth condition – the solid. In this solid state of the rock that formed, we can distinguish older and younger 'igneous' formations of granite and greenstones, and the formation of trachyte and basalt. In the older formations of granite and greenstones we see the remains of the first recapitulation. In the younger formation of the same rocks we have the remains of the second recapitulation, and in the trachytes and basalts the remains of the third recapitulation.

In all these recapitulations the stages progress from heat to fluid. The whole development is closely linked with the development of the human organism. During the third recapitulation, the 'mineral' condenses to the vaporous-fluid-gelatinous condition which Rudolf Steiner described as the Lemurian albuminous atmosphere.

In picturing these progressive solidifications and rarifications one must remember that they also have a spatial aspect. Rudolf Steiner said that, in its pre-physical development, the Earth extended far into planetary space and through a gradual rhythmic shrinking process acquired its present size. During these processes life, which to begin with filled the whole body of the Earth, withdrew more and more to the periphery where there was now an atmosphere in which it could develop further.

Such was the Lemurian albuminous atmosphere. In it were contained all life processes and substances in an undifferentiated state. It was the surrounding which nourished, illuminated and warmed the living Earth. The world-plant-animal, together with the essential being of plant, animal and man, lived in it and constituted the life of the Earth.

The oldest life of this living albuminous atmosphere was the world of the mineral-plants which was brought over from the Old Moon in the recapitulations described above, and now came to an end in a 'grandiose combustion'. This process of combustion described by Rudolf Steiner in the note quoted in

Chapter 2 was, in another context, characterised by him as a sort of flowering process. In order to understand this we have to look at processes which take place in the flowers of present-day plants. What we mean is seen most clearly in annual plants. Let us take plants such as chamomile or marigold (calendula). In their flowers, colouring matter and strong scent are formed. The plant which, before it flowers, is all growth and development, suddenly calls a halt in order to make room for a process which one can only compare with a burning away – a flaming and fading away (disintegrating). What arises from this decomposition are essential oils. These are inflammable and burn with brilliant coloured flames.

It is noteworthy that every blossom has a considerably higher temperature than its surroundings. Here we have really an organic process of combustion. The process does not go as far as mineral combustion. We smell no smoke, only the scent of the flower, we see no flame, only the colour of the flowers, we see no ash, instead we see the seeds which result from this remarkable plant-combustion.

The life of the mineral-plant was a pure blossoming process – what Rudolf Steiner in his note called 'the mineral being consumed in sulphur'. No leaves nor roots belong to this blossoming process. The mineral-plant was an organism that floated in the living, nutritive substance of the albuminous atmosphere and was open to the forces of the surrounding cosmos and the Sun. Into this life of the mineral-plant there dripped and streamed down from the surrounding spheres of the Earth what later densified as quartz and silicious minerals. It caused this life to densify into flowing forms which came and then went again, allowing the more or less formed silicious substances to fall out. These processes were connected with light phenomena – with a glittering and 'greening' which formed and dissolved. Rudolf Steiner described this life of the mineral-plant in his lectures *Mystery Knowledge and Mystery Centres* where he showed that the knowledge of this life on the Earth in times long past was the content and teaching in the pre-Christian temples and mystery centres in Greece and Ireland. He also pointed out that in present-day rocks there are forms which are something like memories of these old times. He said:

> Anyone who has looked round a little on nature knows quite well that something like distinctive markers of an ancient time are to be found in the mineral world today. We find stones, we take them into our hands, we look attentively at them, and we find they have within them something like the outline of a plant-form.★

Before we go into this any further and describe the relevant phenomena in the rocks, we would mention that Rudolf Steiner, in very early lectures in 1907 and later in his lectures to the workmen at the Goetheanum and at teachers' meetings in the Waldorf School, pointed out that the minerals – quartz, mica and feldspar which make up granite and gneiss – are to be seen as the remnants of the flowering process of these mineral-plants. The laminate structure of glittering mica suggests bracts; and feldspar, pistil-like forms. Quartz is to be seen as the filling between these forms. Hornblende which appears in these old rocks is also to be seen as a metamorphosis of these bract-like forms.

As Rudolf Steiner spoke of these 'markers' of ancient times we might add that mica is often found in rosette-like aggregates – 'mica roses' (Plate 5). This is an unusual arrangement of the individual mica crystals which normally form small hexagonal prisms.

The author has found such mica rosettes in the Fichtel Mountains and the Black Forest. They are sometimes embedded in the solid rock but can also appear in clefts and cavities as free standing groups of crystals. The latter especially give a strong impression of a plantlike, calyx-like grouping.

★ Lecture of Dec 1, 1923.

In Moravia (Heřmanov in the Vysočina region) ball structures (Glimmerkugel) are found in which brown mica alternates with fibrous actinolite and nephrite (jade). Also found in Moravia are green flaky prisms of zinnwaldite (lithium mica) with a sharp hexagonal cleavage which are crowned with flaky rays of crystals. Spherical-curved, laminated mica is to be found near Skogböle in southern Finland. From these few examples, that were described in old classical textbooks, we can see what is meant. There may be many more forms of mica which are reminiscent of flower bracts. But these phenomena were considered accidental, they were ignored as unimportant and were not described. It is a task for the future to collect and describe such phenomena which Rudolf Steiner called 'markers' so that we may obtain confirmation for the findings of anthroposophy.

These 'markers' are even more evident in the minerals in greenstones and schists. The principle minerals are augite (pyroxine) and hornblende (amphibole). These illustrate that stage of the mineral-plant development which we described as wood and treelike, which indicates the repetition of the Old Sun condition.

At the end of this development which leads from the flower-like to the wood and treelike mineral-plant world, we have serpentine which, in its duality of foliaceous and fibrous serpentine, forms the bridge between the laminated mica and the fibrous (long, needle-like) minerals of the pyroxine and amphibole groups.

This transition follows two great series of rocks; from granite through gneiss to crystalline schists (which include part of the schistose greenstones) on the one hand; and from granite through syenite and diorite to diabase. At the end of both series we have serpentine which, as foliated serpentine (antigorite), leads through chlorite to mica, and as fibrous serpentine (chrysotile), (Plate 8) to hornblende.

In these groups of rock which occur in nature in all variations together, we see the development in impressive metamorphosis from the granular form-principle of granite to the foliated and fibrous form-principle which appears later in the plant world. The point of departure of this development is mica – this pale image of the primal leaf which has been left behind as a marker of the beginning of life.

We shall see later that this original formative principle of mica is still active today not only in the creation of the plant world but also in the formation of the living, plant-bearing soil.

The threefold repetitions of previous conditions on their long path through many metamorphoses, developing the foliaceous and fibrous form-principles, have left behind a multitude of forms which bear witness to these metamorphoses. From the relative simplicity of granite, a wealth of rocks and minerals has developed which is comparable to the richness of our present plant world.

In the widespread wood-grained structures of crystalline schists and greenstones, which appear embedded in the Earth like annual rings of giant trees, there are innumerable little 'markers' hidden away. The most obvious is asbestos (Plate 9) this densely fibrous mineral that can be spun and woven like plant fibres. We find bissolith, mountain flax, mountain leather and mountain cork – all minerals related to asbestos – minerals which are pliable like leather and which float on water like cork.

Then there is chrysotile – green, densely fibrous layers embedded in gabbro-greenstone, the brown glittering fibres of bronzite, the green diallage and diopside, green pistacite and nephrite, all of which when embedded in steatite (soapstone), schist or quartz, form grass and plantlike structures. On the cleavage surfaces of mica schists one finds delicate crystals of nephrite and hornblende in enchanting forms of fans and sprays (Garbenschiefer) (Plate 7) and patterns resembling leaf veins.

Perhaps the most beautiful of what Rudolf Steiner called 'markers' is to be found in the exquisite green seaweed forms of the moss-agate (Plate 10). Here the bluish translucent threads of hornblende-asbestos or the micaceous chlorite are embedded in a gel-like chalcedony. If one cuts this mineral into thin slices only a few millimetres thick and illuminates them from behind, one seems to see green plant forms floating like seaweed in the ocean.

It is remarkable that most of the minerals that we have called markers are green. When we look into the substances of these markers it becomes apparent that the plant-nature of their origin is also expressed. The main constituent of all the above minerals including their parent rocks (crystalline schists, greenstones, gabbro, serpentine, etc.) is magnesium silicate, and the green colour is caused by iron silicate. So we have three substances which are essential for our present-day plant world. First magnesium which, in the chlorophyll of green plants, takes the place of iron in the haemoglobin of human blood; then iron itself that must be present in the soil and air surrounding the plant so that the chlorophyll in the plant can be produced. And finally silica which not only enables the plant to absorb the light but is built into its rigid form and structure.

From the above it is clear that the old mineral-plant world which originated in the past life-stages of the Earth, still had within it the essence of present minerals. In order that the living plant could develop further as an entity and rise to higher stages of existence, this life had to die. Its corpse we find in a wealth of remarkable rocks and minerals which as markers still speak to us in a language we can understand.

4. The Character of the World-Plant-Animal

The concept of the plant-animal is something which to our modern way of thinking is as evasive as the concept of the mineral-plant which we attempted to outline in the previous chapter. If we want to get closer to this concept we have to remember that Rudolf Steiner himself characterised this all-embracing, undifferentiated state of development of life as world-plant-animal. This suggests a life process that encompasses the whole Earth, and not differentiated kingdoms of nature existing side by side.

We already mentioned that this life process reached its climax in the Old Moon evolution of the Earth and then reappeared in a modified form during the recapitulations at the beginning of the present Earth development.

What really happened in this life process of the plant-animal cannot be put into modern scientific concepts. This is due to the fact that present-day nature with its differentiated forms simply did not exist.

One might be prompted to ask: What has become of the plant-animal in the course of evolution? What sort of signature has it left behind and where are we to look for the present-day forms of life which have developed from it?

In the mineral-plant we could recognise the two life processes which we find today in blossoms and in trees. We showed that from the granular form principle of granite, foliaceous and fibrous ones developed as found in foliaceous and fibrous serpentine. If we follow the corresponding transformation of the plant-animal, it becomes far more complicated. If we want to get anywhere near the right ideas, we must picture the general connection of the present-day plant-world with the insect-world on the one hand, and on the other hand with the living soil in which it grows. One must not forget that there is not only an interdependence between insect-world and flower, as well as other parts of the plant which are above the earth's surface, but that the larvae of a great many insects are equally necessary for the proper life of the soil in which the plant roots live.

If one can view the interaction and interdependence of the insect-world, the plant-world and the soil as a living unity, one can say: this is what has developed in the course of time from the old world-plant-animal.

The interaction of these three realms which embrace the whole Earth is now, as formerly, the foundation of all animal and human existence and has its root and origin in the world-plant-animal. All the phenomena of what is called symbiosis in the plant and lower animal kingdoms – in earth, water and air, even the intestinal flora and fauna in man and animal – have their origin in the world-plant-animal. They are its transformed remnants and traces.

Here it is possible to add something which still has to do with the concept mineral-plant. Until now we have considered those forms which have resulted from the old life processes of the mineral-plant and the plant-animal. For the former these are the flowering processes of the higher plant and the trunk and root formation of our trees. For the latter it is what we have described above as the living connections of insects, plants and soil. There also are the rocks which must be considered as something that has been cast off and left behind. There is though, in every development, a third possibility; the old life processes may be slightly transformed and projected into the present. A short summary can make this clear.

The mineral-plant
- has *left behind as rocks* the granites, greenstones, schists, etc.;
- has *developed further* into trees and the flowers of higher plants.

The plant-animal:
- has *left behind as rocks* the porphyries and such like;
- has *developed further* into the great living community of insects, plants and soil.

The third possibility lies between these two. Thus it has to be a process lying between dead rock and the life processes of plant and insect. We must seek it in the soil where the plant roots are and the insects spend their larval stage.

What takes place within these really very obscure life processes of humus formation deserves to be considered as yet another kingdom of nature, and although this subterranean realm is closely linked with mineral, plant and insect, yet it has its own laws of life. They are the primeval laws of the plant-animal that belonged to the Old Moon condition of the Earth and which, transformed through recapitulations, has now become the soil – the foundation of life. Our present-day soil corresponds to the living atmosphere of the 'Moon-Earth' in the early periods of our present Earth.

If we consider the formation of humus, then three fundamental processes become obvious. The first one, apparently entirely mineral, takes place in the weathering of rocks. The second process takes place in the dead remains of plants and animals. This second process is brought about through the agency of lower plant and animal organisms such as algae, bacteria, fungi and protozoa. The third process is the unification and compounding of the results of the first and second processes through insect larvae (including the many insects which retain larval forms throughout their life such as wood-lice and centipedes) and earthworms.

The first, the mineral process, is to begin with a dissolution and disintegration. The constituents of the rocks – quartz, mica, feldspar, hornblende – decompose through weathering by water, air, heat and cold into their component parts. The resultant end-products are finely dispersed silica, aluminium oxide, magnesium and iron oxides, and the alkaline salts of calcium, sodium and potash. Through the action of the all-pervading water these substances, especially silica, aluminium, magnesium and iron oxides, become colloids which are extraordinarily unstable and easily revert to the crystalline state.

Under certain conditions found today almost exclusively in virgin forests or steppe-soils, there now begins the reconstituting part of this apparently mineral process. From the above mentioned colloidal (gelatinous) substances quite new minerals emerge. Under these apparently entirely mineral conditions only mineral salts ought to be formed, yet the remarkable fact is that very complex compounds of these substances occur – the secondary clay minerals. These elaborately formed, secondary clay minerals have the inner (molecular) structure of mica. But these mica-like minerals are not crystalline like true mica but are colloidal like the original substances resulting from weathering. This means primarily that they have an extraordinary capacity to retain water.

The reconstituting of minerals resulting in entirely new ones (of gelatinous, mica structure) from the products of weathering, is a true life process of the Earth because it does not follow ordinary chemical laws (if so the result would be salts) but the laws of the life of the mineral-plant world. This is indicated by the fact that mica-like and therefore foliaceous structures arise, and that besides clay and silica, magnesium and iron play a major role in the formation of these new minerals.

Thus we have attempted to sketch the third aspect of the mineral-plant existing today as a living process between what has been left behind as rocks and what has developed further as flower and tree formations.

The other aspect of the life of the mineral-plant is the second fundamental process of humus formation described above. It is the transformation of plant and animal residues in the organic part of humus. This organic part of humus which gives our soils their dark colour is a remarkable living substance. It is closely related to tannin which is found in the bark and wood of our trees and in the roots of many plants. If one examines the chemical constitution of this substance, it is interesting to find that it is also related to certain substances found in the scents and essential oils of many plants. Remarkably in both tannin and humus the transformed essence of the flowering process is to be found. This fact throws light on Rudolf Steiner's observation that the life of the mineral-plant was a 'blossoming' or 'blooming' life. However this blooming is under the surface of the earth, therefore it is not essential oils that are produced, but the organic part of humus. A whole host of minute organisms is engaged in producing the dark humus out of the remains of dead plants, wood and cellulose, as well as plant and animal proteins. There is a very well-organised division of labour between algae, bacteria and fungi receiving and handing on substances at the appropriate stages so that finally, out of the most diverse plant and animal remains, there emerges a uniform humus.

In recent decades all these processes have been thoroughly investigated especially by Russian research workers. German scientists like Willi Laatsch and others have taken up these studies and have reached important conclusions concerning the creation and structure of the organic part of humus. Of particular importance is the fact that humus has a resemblance to mica, like the secondary clay minerals previously described, and the physical constitution is colloidal, like the clay minerals.

One cannot overrate the strange fact that here, in miniature in nature, these two processes take place. The one, apparently purely mineral, leads to the formation of new secondary clay minerals. The other, through the digestive process of lower organisms, produces humus. Both substances are composed, like mica, of minute flakes, and are gelatinous – colloidal.

Each of these two substances is, because of its colloidal structure, extraordinarily unstable and very sensitive to changing conditions of moisture, acidity or heat in the soil. Through even slight changes both can loose their colloidal state and become crystalline. In this case the secondary clay minerals would become stone and the humus gradually peat or coal. This means that they would slip out of reach of the life forces, for these can be active and formative only on substances in the colloidal state.

These two processes, the formation of clay minerals and of humus as living, colloidal substances, can be very easily disturbed or even stopped by faulty cultivation – lowering of the water table (drainage), clear felling of woodlands, monocultures and above all by fertilising with chemicals. Salts destroy the colloidal state of the soils because colloids are flockulated by salts. They can then no longer be a vehicle for life forces.

What works on clay minerals and humus while they are in the colloidal stage are the life forces of the plant-animal. It is of the greatest significance that both these substances are eaten by the soil fauna. The organisms are – according to the nature of the soil (woodland, meadow or arable) – wood-lice, mites, springtails, insect larvae and, above all, earthworms.

These creatures devour the earth, they seek places where the two substances are produced. Their life consists in compounding the two substances through their digestive and metabolic processes into stable, ripe humus. They are the great artisans who, in their diminutive intestines, achieve the complete amalgamation of the minute particles of clay minerals and humus acids, combining these with nitrogen, calcium, etc. so that a substance is created that plant roots can take up directly as food.

This end product – stable humus – is still colloidal. It has a real, though simple, life of its own which is expressed in its extraordinary capacity for retaining water and replenishing from the surroundings what the plants have extracted. If these substances in the soil are husbanded they are a hidden treasure which can never fail.

With this we have described the third process in the formation of humus. We have seen that the first stage of these three processes takes place in the purely mineral realm (the secondary clay minerals). The second involves mainly the plant world (the humus acids) and in the third process the 'animal' takes over and creates something new out of the mineral and vegetable material that is neither mineral, vegetable nor animal.

The life of this humus is a synthesis of the mineral-plant and the plant-animal in close conjunction with the individual living organisms in the soil.

Now we can complete our brief summary:

The mineral-plant
1) has *left behind as rocks* the granites, greenstones, schists, etc.;
2) is *active today* in the formation of the secondary clay minerals and organic humus acids;
3) has *developed further* into trees and the flowers of higher plants.

The plant-animal:
1) has *left behind as rocks* the porphyries and such like;
2) is *active today* in the joining together of the building stones of humus (clay minerals and humus acids);
3) has *developed further* into the great living community of insects, plants and soil.

This presentation of the living connection between rocks and the kingdoms of nature as they are today in all their variety, is not only of importance for our study of rocks. If we can really grasp these relationships we have the necessary basis for understanding the practical applications of the agricultural system arising from Rudolf Steiner's researches.

A future agriculture and forestry that does not take into account these relationships will contribute to the decline of the nutritive value of our plants, and our forests and crops will more and more become prey to pests. The pest problem is a revolt by nature against the *disturbed harmony* between soil, plant and insect. The disturbing of this harmony started when fertilising with salts became general practice. The result was the upsetting of the natural life of the soil.

The laws that govern the life of the mineral-plant and plant-animal are fundamental laws of the whole earth and the kingdoms of nature. These laws originated in the distant past of the Earth and, with many transformations, have persisted to the present-day. We have come to know them in the interaction of mineral, plant and insect. In this the insect appears as a guardian of this harmony. The *winged forms* of these real plant-animals, headed by the bees, control this harmony by bringing together *above the earth* the pollen born of the air and Sun with the ovary born of the watery Moon forces. However, the resulting seed can germinate only when, *in* the earth, the *insect larvae* maintain harmony between the Sun-related substances (silica, magnesium) of the clay minerals and the dark, Moon-related substance of humus.

The second process is even more important. It is the *fertilisation of the earth*. What the bee is above the earth, the earthworm is within the earth. The earthworm is king among all those who fructify the earth because it has remained a 'larva', renouncing the chance to become a 'butterfly'. Its red blood, containing iron, shows that it has developed far in advance of related forms. It is *the* representative of the old world-plant-animal because it knows best how to recreate as humus the life of the mineral-plant which has fallen apart into mineral and plants.

Before we proceed to describe those parts and processes of the rocks which are connected with the old life-stages of the Earth let us consider the following.

In the processes connected with humus formation, besides bacteria, fungi and algae (the simplest plant organisms), there is a group of lower animals which plays an important role – the protozoa. These consist of rhizopods with their sub-groups of radiolaria, amoebae, foraminifera, etc. There are also the flagellates some of which, like plants, form chlorophyll or have a cellulose shell. Tripanosoma (sleeping sickness) and spirochaete (syphyllis) which appear in human and animal diseases are flagellates. The last group of these unicellular animalculae is the infusoria. Among these creatures are some who live in the soil and contribute to the making of humus, and some who live as parasites in higher organisms. These lower animals are still extraordinarily plantlike in their metabolism as the appearance of chlorophyll and cellulose shows. Some of them, however, are also closely related to the mineral, for example radiolaria build their shells of silica, and foraminifera of calcium. Here are organisms which form a bridge from the mineral-plant to the plant-animal.

Thus in protozoa we have organisms that clearly point to the mineral-plant and to the plant-animal. They are metamorphoses of those ancient forms of life on Earth. All symbioses (such as between root and root-fungi), which contribute constructively to the great household of nature, are also true metamorphoses of these old forms. Parasites, however, have not followed this harmonious development. They have remained behind and eke out an existence at the expense of higher forms of life.

Another transformation of the world-plant-animal is to be found in the fresh and salt-water animals known as coelanterata. This is a group of animals that even modern zoology calls plant-animals since they are for the most part permanently attached and outwardly resemble plants. To these belong sponges, polyps, jelly-fish and corals. If we contemplate these oldest animals in an aquarium, we are transported back to a time when from the undifferentiated life of the whole Earth the first forms emerged, fleeting and floating, colourful and transparent. In the oceans these first attempts at creation remain for us to see in all their flower-like beauty. They manifest a capacity to produce forms

49

which far outstrip human art (as were beautifully shown in Ernst Haeckel's *Kunstformen der Natur,* 'art forms in nature').

The traces and markers that the life of the world-plant-animal has left in the rocks are very diverse. In order to understand better the nature of these traces, we must take a look at the special characteristics of the animal-nature and how it manifests itself in the lower animals as well as in the embryonic development of higher animals. Primal forms of life are spherical. One has but to think of a seed, a bird's egg or the ova of higher animals which we call eggs in spite of their small size.

In every such ovum or seed the organism returns to the primal beginning, to where the individual forms of nature began to emerge out of the living world-body. From this primal beginning of the seed, the plant-nature develops into the planar surfaces of leaves and the linear formation of the stalks and inner vascular structures. The essence of plant life is the flat leaf from which it derives its nourishment. With the appearance of the flower a new element enters that goes beyond the leaf. What happens is that a number of leaves unite to form a more or less distinct cavity – the ovary – in which the seeds are formed. The creating of such cavities is not a typical plant process. It happens because in flowering and fruit formation something of a higher nature acts on the plant; something of an animal nature. This principle is visible in the relation of flower to insect. Flower and insect belong together because they both originate in the plant-animal.

This animal or astral principle expresses itself in the animal world through the forming of cavities in the body, such as bladder, stomach, uterus, etc. In the embryonic development of animals the formation of cavities by invagination can be followed in all its stages. The characteristic of the animal is its capacity to form organs, which are cavities, by invagination and evagination. By this, certain processes of the environment are drawn into the interior of the body. While plants are able to take in many substances from the air through the surfaces of their leaves, animals have drawn an outer surface into their bodies and formed the passage of the stomach and intestines. Only by means of these 'in-folded leaves' can it digest and make use of its food.

Another formative principle in animals is expressed in the differentiation of the inner organs as regards structure and function. The inner organs, which are all to a certain extent cavities, are all composed of cells. But these cells are modified into liver cells in the liver, kidney cells in the kidneys, lung cells in the lungs, and so on. All the cells are formed after the same pattern, but their shapes and functions differ considerably in individual organs. By invagination animals create both the cavity *and* the particular inner structure and function of the organ. In this way an embryo, after initial simple growth, differentiates itself into an inner world of organs which in their functions support each other.

This forming of cavities and differentiating (one might even call it individualising) has a counter-picture in the world of rocks. To study it in its first beginnings we must walk over the old mountains of crystalline rock that form the backbone of the Earth. In Europe they are, for instance, the Bohemian Forest, the Bavarian Forest, Fichtel Mountains, Odenwald in Germany and the Massif Central in southern France. In the quarries of these mountains we can often observe that in the uniform granite mass there appear 'veins' and 'nests' in which the individual components of the granite are enlarged. The uniform granulation of the granite composed of millimetre-sized crystals of quartz, mica and feldspar is suddenly interrupted. In the neighbourhood of these streaks and nests the granules become bigger and bigger until they fill in a seam or emerge as free standing crystals in the cavity.

This formation in a crystalline rock is called *pegmatite*. It does not occur in faults and clefts caused by stresses and mountain upheavals, but is found in the solid rock masses. This phenomenon must not be confused with the crystal clefts and caves which occur in younger crystalline rocks such as the Alps. The crystals of pegmatite can vary from the minute to the gigantic and they are not always associated with cavities. The largest pegmatites can yield sheets of mica of more than a metre in diameter, together with enormous blocks of quartz, rose quartz and feldspar. In the Urals there was a feldspar quarry in a pegmatite consisting of one single crystal.

The forming of pegmatite was a process that worked from within on the gelatinous fluid mass of rock-in-the-making. The volume of this gelatinous mass was many hundred times that of the later solidified rock. The formation of enormous crystals was *not* in any way associated with existing cavities but it occured *within* the gelatinous mass. If one finds cavities in which crystals have grown, one must assume that originally these were much larger, and with the solidifying and drying out of the rock they have shrunk tremendously. It is characteristic of dense pegmatite, where huge crystals of quartz, mica and feldspar are packed as closely as in fine-grained granite, that larger forms arise than in the pegmatite veins which snake through the rock.

Besides the crystals which constitute the major part of granite – quartz, mica and feldspar – there appear in the cavities of pegmatite a number of other crystalline minerals. These are mainly semi-precious stones – topaz, beryl, tourmaline – and precious corundums such as saphires and rubies. At first sight it appears that these minerals are only present in the cavities but on closer observation of the surrounding fine-grained granite it is found that these precious minerals are distributed in minute particles in the rock mass. The large crystals of the above stones appear to be sucked into the pegmatite cavities out of the granite mass.

The forming of cavities and the differentiating (individualising) of the cells of the organs associated with the forms of flowers in plants and organs in animals, are two phenomena we have now found in the world of rocks. If we call to mind Rudolf Steiner's remark that the forming of granite goes back to the flowering process of the old mineral-plant world, then these phenomena can be understood as the first *traces* of the influence of an animal (astral) principle, higher than that of the plant. In the 'inwardly proliferating, peat-bog mass' of this mineral-plant world there appeared cavities in which the individualising principle of crystal formation could work. The crystals which developed in the organ-like cavities are the inner sense organs of the Earth.

Next to this pegmatite structure another occurs which is the very opposite. In granite and other granular rocks we find veins filled with a very fine-grained material. If this material resembles very fine-grained, light-coloured granite it is called aplite. If it is dark and rich in iron (hornblende and augite) it is lamprophyre. Aplite is nothing more than a granite almost entirely lacking in mica, and lamprophyre is a rock that consists mainly of feldspar and hornblende as well as augite with practically no quartz.

There is an enormous number of these rocks named after their local incidence and variations, and there is no point in enumerating them all. What matters is to grasp the processes which lead to their formation. If, starting from pegmatite together with aplite and lamprophyre, we can follow the whole process and its extention and branching out into all the crystalline rocks including gneiss and crystalline schists, we can recognise a mighty process which we shall designate as porphyry-forming. The differentiation and endless multiplicity of these rocks is a clear sign that instead of the life-processes that formed the simple granite, here other and more diverse life forces are working. The veins, streaks and seams of these aplites and lamprophyres wind through the solid rock mass like the limbs of a gigantic animal or the branches of a mighty tree. They gradually break up, cleave and transform the primal unity into multiplicity; a multiplicity not only of substance but also of form. It begins with pegmatite and culminates in the amigdaloidal melaphyres, of which more later.

What happened exactly? The two processes, one producing cavities and one individualising certain minerals in these cavities, acted upon the gelatinous mass of what later became rock, and differentiated it in an inwardly proliferating growth process. If these phenomena were looked upon impartially, it would never occur to anyone to imagine that these masses were ever fiery and molten. The immediate impression is of something that has *grown* in the true sense of the word.

The porphyry-process in all its diversity is the expression of the forming-power of the world-plant-animal as Rudolf Steiner called it. It acts on nearly all the crystalline rocks, and its effect is more evident in the *forms* it produces than in the actual substance. What are the characteristics of this porphyry structure or form? Let us call to mind once again the pegmatite with its individual crystals and then both aplite and lamprophyre with their shapeless fine-grained mass. On the one hand we have the strongest formative force, and on the other the beginnings of formlessness. At first both parts of the process run separately. Starting from pegmatite the great granite masses develop towards porphyritic granite in which quartz and mica appear as small crystals and feldspar as larger crystals. From aplite and lamprophyre develop the multitude of subsidiary types too numerous to name. Quartz and feldspar dominate in aplite, feldspar and hornblende in lamprophyre, mica being almost absent. When mica disappears it is always a sign that the plant nature retreats and an animal-plant element takes over in hornblende.

Both streams – commencing from pegmatite and from aplite/lamprophyre – can be described as porphyritic, but true porphyry appears where both streams reunite and form a rock that contains large, well-formed, feldspar crystals embedded in a fine-grained or even glassy ground mass. Individual forms appear in the more or less formless medium. In this true porphyry mica has well-nigh disappeared. Granite has ceased to exist and has been transformed by the porphyry-process. The feldspars – the most conspicuous crystals of the porphyries – are that part of granite which, from Rudolf Steiner's descriptions, we can recognise as the pistil in the flowering process of the mineral-plant. In feldspar, however, the flowering principle approaches so near the 'animal-nature' that, in its chemical composition, it has an alkaline calcareous part. In his Agricultural Course Rudolf Steiner explained that the alkalis – potash, sodium and calcium – are closely related. So these feldspar crystals in porphyry are the sign of the world-plant-animal.

In another line of development which proceeds from granite, globular or spherical forms are enclosed in the pegmatite ground mass (Plates 3 and 4). Continuing along this line we find orbicular diorite, orbicular norite (dark rocks consisting of calcium, feldspar, augite, diopside and olivine) and orbicular gabbro. These spherical nodules which we find embedded in the rock are most singular. The core of the nodules – very evident in the polished cross section – sometimes consists of a foreign substance or more often of crystals or groups of crystals. In such a nodule the feldspar can radiate from the centre and be surrounded by a concentric ring of mica. Light bands of feldspar can alternate with the dark mica. An essentially organic picture emerges from the polished cross section. If we remember that the feldspar was the pistil in the 'flower' of the old mineral-plant-world and mica the calyx, we see in such a cross section the picture of a flower with the pistil (feldspar) in the centre and the calyx (mica) arranged concentrically around it. The rhythmic repetition in light and dark layers show us the phenomenon of 'interpenetrating growth' of which Rudolf Steiner spoke. It is also expressed in the spherical shape. The 'flower' was so to speak completely enclosed and grew within itself.

These orbicular rocks are reminiscent of the later agates. In fact in the subsequent development of these processes, we find amigdaloidal melaphyre – the matrix of the agate.★

Melaphyres are dark coloured rocks which resemble basalt and are composed mainly of minerals closely allied to hornblende and augite, with some feldspar. Mica and quartz are absent. In

★ Orbicular rocks are relatively rare. Orbicular granites occur in Sardinia, Sweden, Finland, Ireland, Chile, South Africa, Zimbabwe, Western Australia, New Zealand and Antarctica, orbicular diorite in Corsica, orbicular norite in Norway and orbicular gabbro in Sweden and California.

the amigdaloidal melaphyre one finds those marvellous creations called amigdaloidal agates. The original almond-shaped cavities are either completely filled with the finest layers of silica, or in the remaining hollow of an incompletely filled cavity there are the most wonderful crystals of amethyst and other minerals. Here we have again the principle of cavity formation which, at the beginning of the porphyry process, we found in pegmatite. If we cut through one of these amigdaloidal agates we can have the impression of an organic form. Fine layers of calcedony and opal alternate as many as seven thousand times within one centimeter. Every almond-shaped cavity has a blow-hole through which the silicious substance was sucked into these hollow organs. Through their silica-fillings in fine layers like annular rings these agates show an image of the *plant* world, and through their organ-like shape an image of *animal* forming forces. It is the last sign and marker of the world-plant-animal.

5. The Plant-Nature and Schists

'In schists,' Rudolf Steiner wrote in his notebook (as we read in Chapter 2) 'the plant-being expires.'* What is meant here by schist is not a single rock but a process which formed those gelatinous mineral substances which have fallen away from life and died. As the porphyry process formed the separate crystals and cavities, so schist shows, in general, a layered structure. If we say that the porphyry process begins in granite with pegmatite, we can say that the schistose process begins with gneiss (Plate 1) which often lies immediately on top of granite. We have here the same phenomenon as in pegmatite. Gneiss consists of the same minerals as the nearby granite – namely quartz, feldspar and mica (or hornblende) – but there is no longer a uniform granular rock, for these minerals begin to separate into layers.

On closer inspection it can be seen that it is mica or hornblende which is responsible for the layering. In gneiss, mica no longer appears in 'books' of tiny individual leaves, but the mica leaves lie singly in thin layers between quartz and feldspar. Hornblende, which in granite forms blunt granular crystals, in gneiss makes sheaves and fans of long-bladed crystals and lies flat in the same direction as the layers in the rock (Plate 7). It is clear that mica and hornblende are the two minerals which govern the structure

---

* The German *Schiefer* covers all schists, slates and shales. The translators have endeavoured throughout to use the most appropriate word since there is no collective English term.

of the whole rock. Quartz and feldspar have to adapt themselves to the lie of these minerals.

We pointed out earlier that mica and hornblende are associated with the petal or leafy formation of the old mineral-plant world. If we can see that these two minerals appear to govern the structure of the layered rock, then we may conclude that the forming forces come mainly from the plant nature. In true mica and hornblende schists and in phyllites this is even clearer.

As with the porphyry process, here we have an extraordinary diversity of rocks, but they all show a schistose and layered structure. Sometimes the porphyry and schistose processes overlap and the outcome is a porphyritic schist or a schistose porphyry.

As the development of the schist/shale series progressed, the process became more and more dominant, so that at a certain epoch large areas of the Earth were covered with its enormously thick layers. In the Silurian and Devonian series slates predominate; but the process no longer produces crystalline schists. Slates are predominantly dark, fine-grained or flaky rocks which can be split into thin sheets. Roofing slates and writing slates are made from them. Slate consists of about 75% aluminium silicate which is flaky like mica. It also contains a great number of other minerals in minute grains and crystals, for the most part they are chemically identical to minerals that are also found in granite or the other older rocks, but their structure is entirely different.

When we survey the global process of slate formation we can wonder how this came about. It is extraordinarily tempting to conclude that these slates originated only by the external circumstances of disintegration of the older rocks and their re-forming from sediment falling out of mighty streams of water. It is contradicted by the simple fact that this slate-process was not a localised phenomenon but continued for enormous periods of time and took place over the whole Earth.

Rudolf Steiner said that in this forming of slate the plant-nature dies. We find two important pieces of evidence in support of this. One is the fact that at the end of the great slate period Coal Measures appear, and the other is that slate shows a decidedly organic structure.

Whoever is familiar with schists and slates must be struck by the endless wealth of forms that remind of the wood of tree-trunks. What we have already noticed as woodlike structure in the greenstones and greenstone minerals, appears again here in gigantic forms. Whole mountains of gneiss or crystalline schist might almost be great masses of petrified wood. Magnificent folds of light and dark layers of quartz, mica, hornblende and feldspar look as if the malleable, soft primal substance of what later became rock, was formed by a sort of rhythmic growth. However, we must not imagine that the present-day rock masses always occupied the same space as they do today. They were far, far greater, were soft and flowing and have gradually collapsed, shrunk and dried up. In the living, widely diffused sphere of the Earth there gradually developed the first lower forms of animal life. We find in the dark roofing-slates crablike animals, the trilobites, now extinct. We find shells, snail-like creatures and fishes. The Devonian (armoured) fishes which have an outer skeleton are especially characteristic of the shales.

Towards the end of *this* shale period, something appears which shows that the whole period is subject to an overruling vegetative forming force – the Coal Measures. We pointed out in Chapter 1 that the formation of coal with its enormous seams cannot be explained merely as the remains of forests. Coal has not arisen out of actually formed plants but out of a prolonged period of shedding and casting off which fundamentally was somewhat similar to the formation of rock. In the Coal Measures there is a visible documentation of the transition of the mineral-plant world to the later higher-plant world.

Here it is not only rock that is precipitated or cast off from the

living processes of the Earth, but organic material which remains
as carbon. We find this carbon as graphite and traces of coal in
shales older than the Coal Measures. It is this carbon together
with traces of iron that often gives to slate its dark colour. In
the Coal Measures we find all the transitional stages from shale
to pure coal. We can see from this that there is a real transition
from rock-forming to coal-forming. The living process behind
it becomes more and more of a purely vegetative nature. It is
comparable with the growth of wood in our trees which forms
rings year after year. What was already visible as texture in gneiss,
crystalline schists and slates now becomes quite evident: the
whole Earth has the tendency to become a tree. In slate, shale
and coal formations we are dealing with a 'tree-stage' of the life
of the Earth.

In discussions with teachers at the Waldorf School in Stuttgart,
Rudolf Steiner emphasised that coal does not originate from
plants of solid substance such as we find on Earth today. What
one finds as leaves, bark, trunks and other plant forms in coal are
only emergent transitory forms which, as they arose, immediately
fell away and became petrified. They are semblances of life
which came about like wind patterns in the sand or ripples on
the sea shore. Perhaps one might say they are the first attempts
in creation to mould elaborate plant forms. What became coal
originated in the sphere of life at the outermost periphery of the
fluid layer of the Earth. The rhythmic character of coal formation
can be recognised by the enormous number of successive seams.
Near Aachen in Germany there are 45, and in the Donetz
basin in Russia as many as 225 seams, one above the other and
separated by other strata. The greater the number of seams, the
thinner they are. The average thickness of coal seams is from
30 to 125 cm (1–6 ft), but there are seams as much as ten metres
(33 ft) thick.

At the close of the Carboniferous period came a time of
mountain upheaval, a period of Earth activity that can be
compared to volcanic activity today. We will return later to

the question of this early volcanism. Mighty movements of the partly solidified Earth caused the breaking up of whole mountain ranges, the debris later hardening as sandstones and conglomerates. But these phenomena which increased in intensity throughout the slate period and then died down, are closely related to what we have designated as the 'tree-stage' of the Earth. In order to understand this we must take a look at the nature of present day trees.

One can appreciate the special character of a tree if one compares an ordinary annual such as borage, with its beautiful blue flowers, with a Norway spruce. Borage is a juicy, hairy herb with a soft, weak stem. It makes a profusion of seeds but the first frost kills it and in the spring the seeds produce new plants which grow and flower quickly.

In contrast Norway spruce takes many years before it flowers, and it needs a further year before the seeds in the cones are ripe. It does not mind the winter, it does not even shed its leaves but remains evergreen. For its existence the spruce is independent of the seasons. Only for growing and flowering does it need the spring and summer. This points towards the spruce, like all conifers, having arisen when there were no seasons as we know them. This means that the Sun and Moon stood then in a different relationship to the Earth. At the time of the origin of the conifers the Earth and Moon were still one cosmic body. It was the period which we have just discussed, for it was during the slate period that the Sun separated from the Earth and this caused the transition that we have described as the Carboniferous period. Of course, the conifers of those days were not the solid forms of today but the first trees gradually began to take shape after this time. They are preserved for us for example in the beautifully coloured, silicified wood and tree-trunks of the Permian formation in Arizona, USA. The soft, pulpy wood was impregnated with the fluid silica and completely transformed into agate.

But let us come back to our present-day trees. In one of his

61

lectures to the workmen at the Goetheanum Rudolf Steiner, when speaking of the nature of trees, described wood as a hardening and drying of what rises as fluid from the roots.* The sap draws up minerals which are deposited as dead matter. In his Agricultural Course he compares the tree with a mound of soil on top of which plants grow. One knows from experience that a pile of soil, raised above the level of the surrounding ground, becomes more fertile owing to the better penetration by air, moisture and heat.

Every gardener makes use of this when building a compost heaps. Certain life processes are initiated which form humus from plant remains and soil. We have already mentioned this. In the tree – which, according to Rudolf Steiner, we should imagine as a 'mound of soil' – this also occurs. In the periphery of the trunk, which is solidified 'fluid soil', bark begins to form, and under the bark the cambium layer. This cambium is a very complicated thing. It arises from the sap which flows down from the leafy crown. In this condition one can recognise it in many trees as gum or resin.

If, however, it remains *in* the tree it becomes cambium which constitutes a sort of amorphous root. From this living layer the wood develops inwards, the skin outwards. This outer skin is however not yet the dead material that we know as bark. What this skin is can best be seen if one goes to a forest in winter and scratches the outer layer of a young sapling which as yet has no dead bark, or even of a mature beech tree which never develops a real, thick bark. Then one discovers that under the skin there is a green layer which, even after leaf-fall, still contains chlorophyll and is alive. This is the true skin that envelopes the whole tree-trunk like a giant leaf. On the outside of this skin is formed the dead mineralised bark. This consists of deeply wrinkled or furrowed woody material containing tannin and certain oils. What is this bark in reality?

* *Mensch und Welt,* lecture of Oct 31, 1923.

We have seen how, from the cambium and skin, there develops wood to the inside and bark to the outside. The cambium can be compared with the root-nature and the green skin with the leaf. But where is the 'flower' of this living mound of soil – this tree-trunk? It is none other than the dying bark. This may appear surprising and hypothetical, but it becomes evident in those barks which develop essential oils, for instance the cinnamon tree. Bark is the result of a process of dying as is the flower of a plant, but this dying process does not lead to scent and colour, instead it leads to denser materials such as tannin, resins and oils. In their chemical composition they closely resemble scents and colours in flowers. The 'dying' of the bark comes about through astral forces working from outside, as they do in the flowering crowns of trees and in ordinary herbaceous plants. Leaves and flowers grow on the twigs of the crown of the living trunk or mound of soil as herbaceous plants grow on the solid ground below. The spherical shape of the crown repeats the rounded surface of the Earth.

What we have here described as the nature of a tree is a true picture of a life process that formerly embraced the whole Earth. There was a time when the Earth was still full of life, when it had a sort of cambium or skin layer out of which developed first the schists and slates and then the Coal Measures. Today we have this in miniature as the formation of wood in a tree-trunk with its annual growth rings.

Once this 'earth-tree' had attained a certain size and solidity, other forces from outside penetrated the hardened crust and by tremendous movements the first mountains were drawn up out of the Earth. (Rudolf Steiner remarked that mountains and volcanic upheavals are not caused by pressure from within but by suction from surrounding forces.) It was in this way that the bark was formed on the earth-tree. Today we can still study the same forces forming miniature mountain ranges on the bark of forest trees. The secret of trees is that they have preserved in miniature an old condition of the whole Earth.

Now we can understand what Rudolf Steiner said in his Agricultural Course about the significance of schists and slates for the higher plants. For the herbaceous plants, these rocks play the same role as does the wood of trees for their own blossoms and leaves. The schists and slates of the Earth are the tree-trunk on which the herbaceous plants grow. Alternatively one can say that because once the Earth was a sort of tree, it became possible for most of our food plants to arise.

On the other hand when we considered the rocks that are associated with the bark-forming of this ancient earth-tree (the now familiar series from granite to porphyry) it became evident that they are responsible for the flowering of herbaceous plants and trees. The flowering of the mineral-plant world and the life of the world-plant-animal had to be turned to stone in order that from the dead body of the earth-tree future flowering plants could arise surrounded by bees and butterflies.

6. The Animal-Nature and Limestone

No other rock in the world shows its origin so clearly as limestone. Consider for example the Carboniferous limestone. In the early days of the Earth's history (Pre-Cambrian) one finds limestones, marbles and dolomites which owe their origin to lower animals such as ctenophora and cnidaria (plant-animals) and molluscs, as well as lower forms of plant life such as some algae which we have characterised earlier as transitional forms between plant and animal.

The period in which these first limestones appear corresponds with the beginning of the schist period. All over the Earth one can observe that, wherever gneiss or crystalline schist follows granite or one of its metamorphoses, there are also to be found limestone and dolomite between these other rocks. Naturally they are accompanied by greenstones and porphyries in their varied forms, for this is the time when the great differentiating of life began.

The traces of life in these older limestones are very sparse. Owing to their softness and delicacy, their forms were destroyed. The gelatinous condition of matter also contributed to this destruction. However, there is no need to search for concrete animal forms. The organic origin of these limestones is clear from the great quantities of carbon dioxide which they contain. Another substance of organic origin that plays a great part in these old limestones is graphite. Although as pure carbon it points more to a vegetative process, one must remember that it was just at this time that the plant-animal began to differentiate into the plant and the animal. There was as yet no clear division between the various life processes and forms, and therefore there is a confusing interpenetration of rocks.

The enormous scale of those processes and formations can be seen in Ontario, Canada. There these crystalline limestones and marbles have a thickness of 15 000 metres (50 000 ft). Between the limestones and the often coloured marbles there are layers of quartz, gneiss and serpentine. The marble itself contains mica, hornblende, serpentine and above all, graphite. The graphite content of the limestone may be between 3 and 10%. There are seams of graphite between one and four metres (3–13 ft) thickness. In this area there appears to be more carbon (as graphite) than was deposited (much later) during the whole Carboniferous period over the entire Earth. If one could visualise how much coal is extracted from the Earth daily and how much more is still lying in the Earth, it might be possible to form an approximate picture of the superabundance of life forces behind the forming of these limestone and graphite masses.

What meets us in these early times as limestone without fossils is the first sign of a mighty process that drew animality down to the Earth with the help of calcium. Rudolf Steiner goes into this process in his lectures *Mystery Knowledge and Mystery Centres* which we mentioned earlier. One must not imagine that this process only began when the 'plant-nature' emerged. The life history of the Earth is not a succession of events that can be followed step by step in the layers of rock like turning the pages of a book. It is more of a mingling, an interpenetration of events. It is, however, quite noticeable that certain life processes clearly predominated at certain times.

From these early times onward, limestone is on the increase while silica rocks decrease. This means that the animal-element emerges and makes itself felt. From early periods onwards through the Cambrian, Silurian and Devonian periods (which are the real slate/shale periods) up to the Coal Measures, we find, between the slates and sandstones, increasing layers of limestone and calcareous rocks. When the 'plant-nature' reached a certain climax in the forming of rocks, calcium took the upper hand. This was after the Bunter Sandstone.* In the Muschelkalk we encounter solid limestone masses which extend over wide areas like massive underground masonry. Where Muschelkalk is worked in large quarries we can study its remarkable structure. It consists of bands of dark bluish limestone of thicknesses up to about half a metre, alternating with narrow, darker bands of marl a few centimetres thick. Vertical jointing gives the impression of masonry. The colouration is due to very finely disseminated iron sulphide and bituminous substances. Between these beds occur thick bands of harder dolomite (calcium magnesium carbonate). This dolomite is often silicified. In some layers one finds banks of algal concretions (produced by blue-green algae) and in others quantities of mussels, terebratulids (ancestors of the lampshells which today live only in the depths of the sea), and sea-lily stem sections. These visible animal remains of the Muschelkalk are very poor in species but very rich in individuals. The name Muschelkalk originates from the fact that these few species, especially shells *(Muschel)*, are present as fossils in huge quantities.

The most remarkable feature of the Muschelkalk is its alternating layers of limestone and marl which make it look like solid masonry. When we realise that the Muschekalk area in Europe originally stretched from the northern Germany to North Africa and from western Spain to the Caspian Sea, one wonders how it was possible that this rhythmic stratification could take place over so large a surface. The thickness of the formation varies from about 10 metres (30 ft) at the edge of the area to about 250 metres (800 ft) in the middle.

The stratification of Muschelkalk resembles that of the schists and shales. But instead of siliceous formations we have here

---

* The German succession is described here; in Britain earlier sandstones are equally important (Trans.).

mainly calcarious rocks alternating with the layers of marl. (Marl is more or less calcareous clay). The calcium in these strata is undoubtedly of animal origin and the bands of marl contain much iron sulphide and bituminous substances which account for the dark colour. This rhythmic stratification is a real puzzle for modern geology and one that cannot be solved by the application of present day physical and chemical laws.

In reply to questions put to him by the workmen at the Goetheanum Rudolf Steiner gave a lecture where he spoke of the time which immediately followed the event of the separation of the Moon from the Earth.* There was no solid rock on the Earth at that time. Everything was there in a fluid and gelatinous state in thick layers. Later it shrank, dried out and solidified as rock. Rudolf Steiner described gigantic forms of life which lived in this viscous earth. These gigantic creations which were so big that one could have superimposed the whole of France or Switzerland on their backs were somewhat oyster-like with a scaly surface resembling the shell of a tortoise. These great animals secreted these shells on their upper surfaces, and left behind a slime as snails do today. However they left no fossil remains.

What is important for us in this picture is the idea of a *life process* that led to the formation of the armour-plated shells of these animals. This process was not localised anywhere in particular but extended over the whole Earth and was part of the life of the Earth.

This forming of a carapace like that of tortoises and oysters is a life process of the Earth as a whole, as was the forming of schists. Just as we cannot consider schist as the remains of trees so we should not consider the regular layering of the Muschelkalk as the remains of giant oysters. Instead, the remarkable structure of the Muschelkalk with its alternating bands of limestone and marl becomes intelligible as an organic process retaining something like an afterimage of the carapaces of these giant animals.

* *From Crystals to Crocodiles,* lecture of Sep 23, 1922.

A present-day oyster shell is built up exactly like Muschelkalk on a small scale. In the oystershell, layers of lime alternate with a remarkable substance called conchioline. It is related to the chitin of the exoskeleton of insects and is composed of albumin and cellulose. Every thin flake of lime in the oyster or snail shell is enclosed in a thin veil of this conchioline. The layers are arranged like roofing tiles overlapping each other. The construction of Muschelkalk is similar except that everything is on a giant scale and through petrification the organic origin of the stratification is obscured.

The remarkable stratification of Muschelkalk is an image of life processes into which inorganic matter subsequently penetrated. It is fundamentally the same phenomenon as in moss agate which we described in the chapter on the plant-animal and its signature. There as here, mineral matter seeped in, leaving a picture of a former life process in the discarded sheath.

What we find in Muschelkalk as actual animal forms – mussels, snails, crinoids, etc. – stems from a later time and lived in the discarded remains of the ancient life.

Thus in Muschelkalk we have found a picture that represents a life process that we might call the 'oyster' stage of the Earth. It was a preparatory stage for the further differentiation of the animal world. The giant forms disappeared and the life of the Earth as a whole dispersed itself into tiny forms which subsequently built up the mighty calcareous layers of the Jurassic and Chalk periods. At the same time there gradually arose the new giant forms of the saurians. This is the beginning of the new transition that leads to the mammal era. After the precipitation of calcium in the Chalk, the 'life' of the Earth begins to fade away. The rocks of the Tertiary period (Atlantis) can only to a very limited extent be considered as new formations of the living Earth. From the middle of this period dissolution set in; the breaking up of what had been formed.

In the foregoing we have endeavoured to sketch the life processes connected with calcium carbonate. It should be noted that this is the form of calcium which is associated with the lower animals and the lower plants (algae). These lower organisms are the first forms of life that have been preserved even in the very oldest strata.

In the aforementioned lectures, *Mystery Knowledge and Mystery Centres,* Rudolf Steiner gave a vivid description of how these first animal forms arose in the soft fluidity of the Earth. In the rarefied, watery albuminous atmosphere existed not only vegetative processes, but also the first attempts at animal forms, as transitory and fleeting as the plant forms. He speaks of how lime had the capacity to evaporate like water after being incorporated in the fluid Earth. There was a rhythmic up and down movement of lime as vapour and rain. As the lime descended and densified it enwrapped the soft animal forms and drew them down to earth.

Thus these first tender, soft, fluid forms were clothed with shells and protective armour and appeared on Earth in more solid forms. As molluscs and bony-plated fish we find them in the old strata. Many molluscs remain little changed to this day but the bony-plated fish have died out. They were an early form of a group which has since developed further.

Though Rudolf Steiner's description of the evaporating and precipitating lime may appear surprising, yet lime still has this property today. We have to remember that this process did not take place in an atmosphere such as ours, but in a 'fluid' atmosphere. Naturally lime can no longer clothe fleeting animal forms and bring them down to the earth, but it still has the peculiar capacity, in lower temperatures with the help of carbon dioxide in the air, to dissolve in water and become bicarbonate of calcium and, when this water is heated, to precipitate as solid lime. This is a process continually taking place in nature which plays an extremely important role in the maintenance of life in plants and animals. If lime were as insoluble as silica, no living organism could take up this substance which is so important both for support and nutrition.

There is another form of calcium in rocks which quantitatively cannot compare with calcium carbonate. It is calcium phosphate. Whereas calcium carbonate is mainly the result of the living processes of lower organisms, calcium phosphate is the skeletal substance of vertebrates and of man.

In rocks calcium phosphate appears in two forms. First as well-known apatite (natural calcium phosphate plus either fluorine or chlorine) and as phosphorite. Apatite is a component of pegmatite and many of the greenstones. One can also find it in younger basalt and other lavas. On the other hand phosphorite has formed in sedimentary rocks (sandstones, shales, marl and limestones) from bones and proteins of dead animals. It is unmistakably of organic origin.

Mineral apatite, which often appears in wonderful crystals in the granular igneous rocks, has exactly the same chemical composition as the substance in animal and human bone. This is a remarkable fact, for here higher animals and man have incorporated in their bodies a very clearly-defined *mineral* process. One can recognise the crystalline structure of apatite in a thin section of bone. It is present quite independently of the bone cells.

Now we know of two metals in particular which are associated with apatite – iron and tin. The enormous deposits of magnetic iron ore found in northern Europe and northern Canada are completely permeated with large quantities of apatite. And in this apatite are to be found minute quantities of lead phosphate.

If one looks at human bones the way Rudolf Steiner did in his medical lectures, one finds that these three metals – iron, tin and lead – play an important part in the forming and functioning of the bone system.

From the interior of the bone iron finds its way into newly created blood. In the joints which bring movement into the rigid skeleton – and which are filled with joint fluid and thus maintain a balance between fluid and solid – tin is active. (This does not mean that tin is present as a substance). In the hardening of the bone, lead is active. Chronic lead poisoning produces calcification of the blood vessels and, in certain circumstances, an excessive hardening of the bones. In such cases lead is deposited in the bones. However in healthy, normal bones one does not find lead.

In apatite we have another example of how an organic process – in this case bone formation and function – is the prototype for phenomena in the apparently lifeless mineral kingdom which otherwise we would find strange and puzzling.

7. The Human Being and Salt

Up to this point we have surveyed in broad lines a major part of the rocks of our Earth and have tried to show how these now dead formations have in the past emerged from life processes of the whole Earth. We found that these once all-embracing processes appear today transformed as certain phenomena and forms of life in the various kingdoms of nature.

When Rudolf Steiner linked the human being – who has cast off and left behind the animals, plant-animals, plants and minerals – with the forming of salt, it is not as easy to see the connection as with the other rock formations. If we wish to understand this connection, we must turn our attention to the significance of salt in human life processes: not only those contributing to nutrition, growth and regenerative powers, but also those working in consciousness, imagination and thinking.

In order to appreciate what far-reaching significance salt has for the whole human constitution, one has only to deprive a healthy person of the salt normally used for seasoning. The first symptom that appears is loss of appetite. This is due to the fact that salt has the peculiarity of bringing out the individual tastes of all foods and, through the experience of taste in the area of mouth and tongue, it works deeply into the unconscious functions of the inner glands connected with the stomach and intestinal digestion. Thus, through salt, something becomes conscious that works unconsciously in digestion and nutrition. The *inner* taste is stimulated which, according to Rudolf Steiner, extends right into the liver. This *inner* taste, which only appears consciously as appetite, is linked with another peculiarity of salt. It guides individual nutritive substances to the appropriate places in the organism.

The reduced fluid requirement of a person on a salt-less diet shows that the organism needs to retain the salt present in the body fluids. Normally the salt in the blood and tissue fluids would be continually exchanged. In this process however, which regularly introduces new salt into the system, lies one of the most important biological functions connected with digestion and nutrition.

The withdrawal of salt has another significant effect. We become tired and apathetic and find that our ability to think is impaired. These phenomena are all connected with the other side of our life; with our consciousness. The nature of these two processes is diametrically opposite. Digestion and nutrition build up, while consciousness arises as a result of a breaking-down process. Physiologically, consciousness and self-awareness destroy what is built up and this breaking-down takes place almost entirely in our nerves and brain. In this nerve-sense system, as it is described by Rudolf Steiner, human beings are able to concentrate nutritional substances to the point of mineral density and lifelessness, but are also able to dissolve them again, destroy and cast them out from their organisms. These physiological 'death-processes' are specifically human, and do not occur in animals.

Ancient terminology for this was 'salt-process'. It signified that out of something in solution a hard substance was precipitated. In this sense one could even call the forming of all rocks and minerals of the Earth a salt-process. But one must note that most rocks and minerals are *not* soluble in water – at any rate not to the extent of real salts. Only true salts are soluble in water.

In humans we find salt in two conditions; first in solution in the blood, and then as *process* in the brain and nerve system which leads to a sort of depositing, and which is the physiological basis of consciousness. The best-known example of this is brain-sand (corpora arenacea), the calcified structures in the pineal gland which, if it is absent or abnormal, may cause mental illness.

Where salt is present in solution as in the blood and digestive juices of man and animal, we have the primal condition of salt as we find it in nature in sea water. The oceans of the world are but the remains of that albuminous atmosphere of the Earth which we have described earlier. In this amniotic fluid of the living Earth the living forms of the kingdoms of nature developed. They cast out the mineral part into the mineral world and drew into their systems a part of this living atmosphere as blood and the other bodily fluids. In this way the original life of the Earth itself was divided up into individual forms in the kingdoms of nature. When the old albuminous atmosphere gradually decomposed, there arose out of it the present-day oceans and the atmosphere. When albumen (protein) decomposes there appear the substances present in seawater and in air, namely: salts, water, oxygen, carbon dioxide and nitrogen.

The oceans are still today a real amniotic fluid for a multitude of organisms. It is a physiological fluid in which the oldest and most primitive forms of life could persist till the present day. How near to life the substance of sea water is can be seen from the proportions of soluble salts in it – especially sodium, potash, magnesium and calcium salts. These proportions resemble those of human and animal blood. Pure sea water, diluted to the salt level of human blood, could be mixed with medicinal extracts of plant or animal origin for injection into the blood stream.

In contrast with the primal condition of salt in solution, still closely connected with life, we must also consider quite a different process which has deposited salt in the Earth in solid crystalline layers like rock.

The enormous beds of rock salt in the depths of the Earth are distributed over all continents and are found from early Cambrian and Silurian times up to the Tertiary system.

The beginning of this deposition of salt in the Earth coincides more or less with the differentiation of living things and of rocks that we spoke of in connection with slates. It is the time when limestone and dolomite began to appear. The deposition continued through the subsequent formations reaching a first *maximum* immediately after the Coal Measures in the Permian system. The potash and rock salt beds of Stassfurt in Saxony-Anhalt belong to this system.

A further maximum in salt formation occurred in the Tertiary system when the forming of new rocks and strata from original life processes was nearing its end. The potash beds of the Upper Rhine basin, Poland and Spain belong to this period. When considering the great quantity and widespread occurence of rock salt one has to bear in mind that salt is a soluble substance which has taken on the appearance of rock. But we must also point out that though salt today is soluble, it does not mean that these huge beds were formed from a fluid solution such as sea water. Rudolf Steiner remarked that the solubility of minerals which we observe in salt is the most recent characteristic that has developed in the mineral kingdom.

This observation becomes intelligible when we bear in mind that all rock was originally in a gelatinous colloidal condition. Gels and colloids are original states of matter, not solutions. They have a close kinship with living matter – protein.

The whole appearance of the salt deposits suggests that this salt-rock, like the slates and coal measures, has been cast off from a major life process of the Earth. Rock salt occurs in layers and – as for instance in the Permian (Zechstein) salts – frequently alternates with potash and magnesium salts interspersed with thinner or thicker layers of anhydrous gypsum. The salt deposits are almost everywhere encased by massive layers of anhydrites which are, as often as not, accompanied by dolomite. These enveloping rocks are impermiable so that neither surface nor underground water can normally penetrate to the salt beds. So we have the curious fact that a soluble substance in the interior of the Earth is thus protected from contact with water.

One can imagine what it would mean if all the salt buried in the land masses were exposed to the influence of water, since the quantity of rock salt equals approximately the amount of salt in solution in the oceans – thousands of billions of tons. Since about 70% of the Earth's surface is ocean and barely 30% is land, the rock salt, if it were not protected, would permeate the soil with a salt content rendering all life impossible.*

The fact that this does not occur points to a rational organic process in the forming of salt beds. This process is connected with the human being, in the same way as is slate with the plant, and limestone with the animal.

In Rudolf Steiner's note mentioned in Chapter 2 he wrote 'in salt the universal human being is extinguished'. This means that the first appearance of salt is connected with the fact that the universal human being, until then part of the Earth as a whole, began to differentiate into individual human beings. In order that individual human beings could develop further out of the universal human being, the animal nature, plant nature and plant-animal had to die away, and the mineral be burnt up in sulphur. At the same time the differentiation began of the kingdoms of nature, which is reflected so strikingly in the diversity of the rocks. Together with porphyry, schists, slates and limestone, salt also appeared early as an expression of the fact that now a being began to act that had the capacity to overcome the hardening of matter through the dissolving power of living water. Human beings are the only beings able to make use physiologically of the mineralising decomposition in the nerve and brain system in order to develop consciousness of selfhood and power of thought. The reflection of this human capacity in the world outside is the ordering of salt deposits among rocks. Before human beings completely descended to the Earth they condensed 'thoughts in salt'. Through the power of thought the human being banished the death forces of salt into the Earth. This salt was in a condition that had nothing to do with its present solubility. It was still a colloid that obeyed organic laws.

---

* The salt and mineral content of the oceans alone would cover the surface of the Earth with a layer 45 metres (150 ft) thick.

The concept that a substance which today is highly soluble in water was once a paste-like colloid that swelled as it absorbed water but did not dissolve, is certainly very unusual. But there are some phenomena that justify this idea.

In many salt deposits, especially in rock salt, we find, between layers of coarse salt, lenses or layers several metres thick of clear salt that has almost glasslike transparency, and is hardly streaked or clouded at all. If a light is put behind such a block in a mine, the whole block is illuminated like a vitrious mass and the source of light can be seen through the clear salt. If the salt had crystallised in lagoons and shallow bays under the influence of tropical desert temperatures it could never have this clear structure, but would be deposited as a crumbly mass of small crystals.

There is another phenomenon which points to the malleable-colloidal condition of salt. In 1937 when test drilling for oil in the Hanover-Celle of Germany region, a boring was made which entered a salt dome at a depth of 475 metres (1550 ft) and penetrated to 3 818 metres (12 500 ft) without reaching its bottom. At that depth the salt had a temperature of 130°C and was a doughlike malleable mass in which the drill got stuck. Since the melting point of rock salt is over 800°C this is a strange phenomenon which shows something of the state of salt in the depth of the Earth. In this connection it is also interesting that, when making the Simplon Tunnel through the Alps, crystalline quartz was found in the heart of the mountain together with a gelatinous colloidal silicon dioxide – a silica that was still in a similar primal state (before crystallisation).

With the discovery of this salt dome over three kilometres thick we can gain an impression of the tremendous size of these salt beds. Hundreds of such salt borings were made in northern Germany, and we now know that the Zechstein salt beds alone cover an area of about 100 000 sq km (40 000 sq miles). Potash and magnesium are extracted from about 250 pits and converted into fertilisers and chemicals. Only the upper horizon of these

Zechstein rock salt beds contain the valuable potash salts; the beds lie at depths of some hundreds to over a thousand metres. Below them usually lie still older layers of rock salt which are not worked since there is enough pure salt more easily accessible elsewhere.

In Europe the Zechstein salt beds stretch from England in the west eastward to Poland, and further east still lie the great salt beds of the Urals.

The Zechstein salts constitute only the older salt beds of this area. At higher horizons there are the salts of the Bunter Sandstone and Muschelkalk. In Germany the latter extend from the district of Magdeburg through Thuringia, Lower Franconia, Heilbronn and the eastern Black Forest to the Swiss border.

Very young Tertiary potash and rock salts occur in the Upper Rhine basin between Basle and Freiburg at a depth of 500 to 1000 metres (1650–3300 ft). Enormous deposits of salt are found in Russia especially at Kazakhstan where 1500 to 1800 individual occurrences lie only a few hundred metres below the surface but extend to depths of 5000 metres (16400 ft). The Dossor one is 12 km long, 8 km wide and 5 km deep (7.5, 5 and 3 miles). The Iskine one occupies an area of 30 sq km (12 sq miles) and is one and a half km thick in the south and about 5 km in the north (1–3 miles). Another near Osinki covers 70 sq km (30 sq miles).

Through Russian Central Asia there are thousands upon thousands of larger and smaller salt occurrences. They link up with the salt beds of Turkey, Syria, Iraq and Jordan. The last includes the Dead Sea with its forty thousand million tons of salt in solution, one of the strangest and most remarkable salt phenomena of the Earth. One of the richest areas is Iran and the islands of the Persian Gulf. On these there are open cast salt mines or rather mountains 5 to 10 km (3–6 miles) thick with furrows of salt glaciers ploughing through them. Pakistan is famous for the Salt Range east of the Indus where very old salt lies in enormous quantities to a great depth.

The great salt beds of North America are to be found especially in New Mexico and Texas where deep borings for oil revealed beds of salt 300 to 500 metres (1000–1600 ft) thick over an area of 200 000 sq km (80 000 sq miles).

In the course of prospecting for oil in the Gulf of Mexico, hundreds of salt domes were encountered which were so thick that drilling was abandoned. The Five Islands salt domes of southern Louisiana are famous. They tower over the plain and extend underground to a depth of several hundred metres.

We have mentioned only the most important salt formations which can perhaps give some idea of the tremendous and remarkable phenomena of the salt rocks of the Earth.

8. The Mystery of Oil

Having dealt with the mineral-plant and plant-animal stages we must now interject something else that is inseparable from the Earth's life – the origin of oil and its related substances such as asphalt, bitumin and mineral wax.

Numerous theories have been put forward to explain the appearance of these substances in enormous quantities in particular regions and during certain geological periods. One of the first theories attempted to attribute the origin of oil to purely inorganic processes such as can be produced in the laboratory from carbides (metal compounds of carbon) which, in contact with water, become hydrocarbons. Such hydrocarbons are for instance methane (marsh gas) or acetylene, which arises when water causes the decomposition of calcium carbide. But since oil consists of far more complicated compounds, this theory was soon dropped. Besides, in the meantime, it had been shown that the formation of complicated compounds from these simple hydrocarbons could only take place at high temperatures and under high pressure, and high temperatures were out of the question.

Later it was assumed that oil was formed by a sort of distillation of organic remains such as coal and peat through volcanic action in the interior of the Earth. This theory too had to be discarded when it became ever clearer as new oil fields were discovered that oil is never found far from the place of its origin and no volcanic rocks are ever present in the vicinity. It is a peculiarity of oil that it does not appear until a certain period and then exclusively in sedimentary rocks such as sandstones, sands, shales, marls and limestones. It has never been found in metamorphic rocks which are mostly distinctly crystalline. At the present time it is generally accepted that oil is of organic origin. It can be traced back to a transformation of plant and animal remains whose proteins and fats, under the influence of pressure and mild heat throughout long periods of time, became present-day oil. To these factors the latest theories add the activity of certain bacteria which have been found in the salty water of oil wells.

These oil bacteria have definite pecularities. It has been known since the end of the nineteenth century that there are bacterial organisms which can create simple hydrocarbons such as marsh gas out of organic remains under anaerobic conditions. The metabolism of these anaerobic bacteria does not need the free oxygen of the atmosphere because it is able to take this vital substance out of organic remains such as lignin, protein or fats (which all contain oxygen in some form). The organic matter is thus broken down into hydrocarbons and water. In contrast, aerobic bacteria need atmospheric oxygen to live, and give off carbon dioxide. Our yeasts which 'ferment' sugar into alcohol, carbon dioxide and water, work in a similar way.

During the first decades of the twentieth century bacteria were discovered which could 'digest' rubber and some volatile hydrocarbons such as benzol, toluene (phenylmethane), etc. It was then noticed that certain anaerobic bacteria remained alive in hydrocarbons and did not die off. In the middle of the 1920s bacteria were first found in the oil wells and their salty water. These bacteria came from a depth of many thousand metres and must have existed in the interior of the Earth for countless millenia. They have remained alive in salt solutions of varying concentrations which were found with the oil. Neither the high pressures presumed to prevail at these depths nor the relatively high temperatures had destroyed them.

Under laboratory conditions it was established that in an artificial nutrient, some of these bacteria were able to multiply entirely anaerobically at temperatures below freezing point and others at temperatures of 85°C (185°F).

Many of these bacteria were completely new varieties quite unknown on the surface of the Earth. Some strains were able to break down any organic material into petroleum substances. Others again took these and metabolised them into other simple hydrocarbons. In effect researchers were faced by a plethora of new phenomena for which no common denominator could be found. Furthermore substances were discovered in oil that inhibited the growth of bacteria but did not kill them. Closer examination of these inhibiting substances revealed the surprising fact that they were organo-metallic compounds containing heavy metals such as copper, nickel, iron, molybdenum and vanadium. These compounds closely resemble haemoglobin, chlorophyll and the curious substances found in the blood of molluscs and echinoidea, (haemocyanin which contains copper) and the blood pigment of the sea-urchins (containing vanadium). Also there is a red colouring matter containing molybdenum in the nitrogen bacteria attached as nodules to the roots of leguminous plants which is very similar to the haemoglobin in the blood of higher animals and humans that contains iron.

In the early 1940s when all these phenomena were discovered, a group of American scientists followed up the problem to see where it would lead. During this research the question arose as to why, in the course of millenia, all oil had not been 'devoured' by these oil-digesting bacteria and transformed into simple hydrocarbons such as methane, etc. They tried to explain this by the presence of salt water, the inhibiting substances mentioned, and oil itself.

Scientists consider that the answer to this question has not yet been found. But in fact the answer is provided if we follow what happens when certain methods are employed in boring for oil.

When sinking a well it is necessary to continually flush the drill hole with water so that the fragments of rock loosened by the drill are washed away. When the first traces of oil are found in the water brought to the surface, one has 'struck oil'. In order to wash away the debris more effectively certain substances were often added to the water. It was observed in certain cases in the denser rocks, that after an initial flow of oil it gradually dried up again. Pumping water into the bore hole to force the oil out of the porous rock achieved the opposite: the oil well dried up in spite of the fact that great quantities of oil were present.

After much investigation it was found that the water reawakened the bacterial world to new life and that the products of their metabolism so changed the viscosity of the oil and salt water that the pores of the rock were blocked and the outflow of oil prevented.

This fact is of far-reaching importance for the understanding of bacteria and their activities. It shows that bacteria in oil and its associated waters are now in a state of arrested activity, but this does not mean that they were always inactive. The tremendous quantities of gas present in most oil fields and consisting mainly of methane show quite clearly that bacteria indeed have transformed some of the oil. But this past life process in the depth of the Earth has come to a certain equilibrium. The moment conditions are altered for the quiescent bacteria – by pumping in water – their activity is reawakened.

Thus we find in association with oil an extraordinary rich and varied bacterial life which appears to be in a more or less quiescent state. In oil itself one finds substances which are closely related to certain vital substances familiar to us in nature today. Apart from the organo-metallic compounds, there are also some that have the effect of oestrogens; that is, they act on the sexual organs of higher organisms like hormones. The remaining constituents of oil are compounds which are reminiscent of resins, waxes and essential oils of plants, together with substances containing sulphur and nitrogen more suggestive of animal proteins. If humus-like materials are present, the oils are dark and rich in asphalt (bitumen).

In its crude state, or isolated into single constituents, oil has certain optical properties. A polarised ray of light is rotated when passing through the fluid. This is a pecularity of substances which originate from life processes. Inorganic and synthetic ones do not show it.

Since it is now certain that oil has arisen through life processes, we may ask what sort of a life was it? Was it plantlike or was it more animal? The constituent parts of oil suggest sometimes one, sometimes the other. Perhaps we can get a clearer idea if we can picture, even if only vaguely, the enormous extent of this life from which oil originated.

In the 85 years between 1870 and 1955 two billion tonnes of oil were extracted from the whole Earth. One cannot visualise such quantities! More then 10 times this quantity remains in the ground. These figures only refer to oil sources which flow out naturally or can be pumped out because they lie in sands, sandstones or porous limestone. Besides these there are the vast strata of oil-bearing shales which in the USA alone are estimated to contain 10 billion tonnes. These oil-bearing shales occur all over the Earth but their oil can only be extracted by mining operations since the fine-grained shales hold the oil like a sponge. The quantity of oil contained in these shales is at least as great as the estimated quantities in sand and sandstones.

Apart from all this, there are thousands of oil wells which, besides oil, produce thousands of millions of cubic metres of gas, as well as borings which yield gas only. All these owe their existence to the activity of the oil bacteria. To add to these inconceivable quantities, there is also the thickness of the oil-bearing strata to take into account. They range from 500 to 4000 metres (1 600–13 000 ft).

At the time of writing the deepest borehole in the world, in Plaquemines Parish, Louisiana, struck oil at 6 880 metres (22 570 ft) in 1956. This oil is in relatively young strata. In Kern County, California, there is a borehole which struck oil at a depth of 5 457 metres (17 904 ft).★

★ In 2012 the deepest oil borehole was on the Sakhalin shelf in Eastern Siberia at 12 376 metres (40 604 ft).

Such thicknesses are comparable to those of the old rocks, slates and chalk. If one looks at the geological periods when the oil-bearing strata originated, two main epochs can be distinguished. The first extends from the 'slate' period of the Silurian and Devonian systems to the Coal Measures. Its oil content is estimated at about one third of the world total. The remaining two thirds belong to a second epoch which begins round about the Chalk times and extends far into the Tertiary.

Most of the oil of the first epoch lies in the Midwest of North America while the young Tertiary oils are mainly found in the Eurasian continent with the heaviest concentration in the Persian Gulf. Part of the younger oils lie along a line stretching from Wyoming and California across the Gulf of Mexico to Venezuela.

No remains of larger animals or plants are found in the oil-bearing sands, sandstones, shales and limestones. The sands show no trace of fossil shells. The limestones are also relatively poor in fossils. Only the shales are often built up in their entirety of minute shells of marine or fresh water organisms such as our present-day diatoms or radiolaria.

If oil originated only from these masses of the remains of dead marine organisms, one should find them in *all* oil-bearing rocks. Conversely there are many deposits of similar remains which contain practically no trace of oil.

Because of this, researchers tried to account for oil by the dying of masses of algae, but there are too many substances of animal origin in oil for this hypothesis to hold good. It may nevertheless be true that the dying of masses of minute organisms such as algae, diatoms or radiolaria is partly responsible for the formation of oil. The major part of the oils of both epochs originated in the tidal waves of life which literally bloomed in the warm watery element of the life of the whole Earth as it was then. Let us think back to the flower nature of the mineral-plant and plant-animal of the young Earth which we described in previous chapters.

In oil we have the organic residues of the tremendous

disintegration of a life not yet differentiated into individual animal and plant forms. This life followed its course in atmospheric conditions which today we only find in a densified form in the rivers and oceans. It can happen today that in warm waters a 'flowering' as one might call it, takes place. Suddenly, through certain climatic conditions – warm currents, etc. – a tremendous proliferation of lower organisms occurs which rises like a flood for a short time and then quickly dies again. Today oil no longer appears in any quantity from such 'flowering flooding life', though in tropical regions one sometimes sees a slight film of an oily nature on the surface of water which shimmers with iridescence. The real production of oil belongs to epochs which had far more exuberance of life than is known anywhere on Earth today.

In the lectures, *Mystery Knowledge and Mystery Centres,* previously quoted Rudolf Steiner divided the creation of life into three great periods. From the first no traces of life come down to us, only the oldest rocks of the Earth. From the second we have the old oils (lying mainly in America), coal and the rocks of the slate period (Palaeozoic) with their traces of plant and animal life. From the third period of creation come the younger oils extending from the Cretaceous to the Tertiary systems, the sandstones and limestones of the Earth's 'middle age', and brown coal. In the Cretaceous epoch we find the Saurians, and in the Tertiary the remains of the ancestors of our present day plants and animals.

When, during the Tertiary period – before the beginning of the Atlantean development – the life of the Earth itself began to dissipate, the remaining life of the atmosphere was precipitated in tremendous organic deposits. These are the inconceivably large quantities of the younger Tertiary oils. From their composition they show that they have more of a plant origin. The transformation of these remains of life into oil is still therefore a life process of the Earth as a whole. It is a process that we can still see today when our plants produce essential oils and resins. Substances are then created which have a close resemblance to oil. It is therefore not surprising that in one of his medical lectures Rudolf Steiner said that the Earth was able to create oil and that these oils are somewhat similar to plant oils.

In one of his lectures to the workmen at the Goetheanum Rudolf Steiner spoke of another kind of oil production.* It takes place in the formation of humus and is of importance for the flowering process of plants. We see therefore that this 'flowering flood of life' is still present today though very diminished and transformed. It no longer produces oil as formerly but serves those parts of the plant which are able to produce oils. The oily humus substance mentioned above is also found in the dark bituminous oils and in asphalt itself. So everywhere it can be seen that the life of the Earth has descended from the surrounding atmosphere and come to work in the Earth.

* *From Sunspots to Strawberries,* lecture of Sep 9, 1924.

9. Sedimentary Rocks

Up to this point we have been considering the rocks that were precipitated directly out of living processes of the Earth and which underwent little or no mechanical change before hardening. If we now look at formations which give the impression that they originated entirely through the breaking down of other rocks and the sifting of the particles through the agency of water and air, we must be careful not to fall into the error of attributing all such formations to processes that we see today as weathering, breakdown and precipitation. True sandstones comprise only 0.7% of all rock, which shows how little of the primal rock has been worn down in order to build them.

To begin with we have to get a picture of how sandstones are distributed in time throughout the various geological epochs. The first appearance of sandstone we can place at the beginning of the Schist period. At the same time the first true limestones appear. But neither limestone nor sandstone are characteristic of this period. By the time of the Coal Measures thick limestones appear – the carboniferous limestones – and there are sandstones lying between seams of coal.

The first great sandstone epoch begins after the Carboniferous – in the Permian system.* It consists of the sandstones and conglomerates of Rotliegend formation which adjoin the massive thick beds of Bunter Sandstone. While the rocks of the Rotliegend formation usually have a red colour throughout due to iron oxide which, together with clay or silica, forms a cement

* In this paragraph it is the German succession is described; in Britain earlier sandstones are equally important (Trans.).

or mortar between the sand particles, the Bunter Sandstones are multicoloured – grey, yellowish, greenish or red. This first sandstone epoch is followed by the limestone of the Muschelkalk. In the Keuper formation which follows the Muschelkalk we have a second sandstone epoch which gradually comes to an end in the blacksand of the Jura. Following the reappearance of limestone in the white Jurassic, we find sandstone appearing for the third time in the Cretaceous and Tertiary deposits.

This rhythm between sandstone and the residues of animal-organic processes is a true picture of the far-reaching and ever-changing life processes of the Earth. While in the limestones as well as in the slates and old granitic rocks we have a direct precipitation from a life process which creates the rock, we see for the first time in the sandstones, conglomerates and marls the effect of mechanical forces which break down and disintegrate these old rocks. We should classify as true sedimentary rocks only those which are clearly the result of the mechanical breakdown and grading of various formations. A closer look at sandstones and conglomerates (cemented masses of pebbles) will enable us to better understand the origin of these rocks.

Sandstones consist basically of quartz granules, slightly worn but mostly sharp edged, which are cemented together. The binding material of these grains of sand is either a form of silica, clayey matter, calcium carbonate, dolomite or (more rarely) iron oxide. The binding matter of conglomerates is equally varied. The nature of the binding is not dependent on the age of the formation. The various sandstones – silicious, clayey, calcarieous – can be present from the very oldest epochs up to the Tertiary, and one sometimes finds many kinds of binding within one formation, for instance in Bunter Sandstone.

In almost all sandstones fine flakes of mica are more or less visible. The only exceptions are a few pure silicious sandstones which consist of up to 98% quartz grains bound with silica, the remainder being iron and a little water (quartzites).

The mica flakes and the clayey binding indicate that the rock is the product of the debris of earlier crystalline rock such as granite or gneiss. The great masses of the Permian and Bunter Sandstone and their conglomerates usually lie immediately above the ancient crystalline rocks from which they originate.

The question now arises, what was the nature and hardness of the rocks whose debris formed the sandstones? The sharp-edged grains and scree in the breccia show that their source rocks were relatively hard. Conglomerates have a similar origin but the shape of the grains of sand and the pebbles indicate that sometimes the source material was less hard and more easily worn by the action of water. This phenomenon is a striking confirmation of what Rudolf Steiner described as the condition of the Earth shortly before and shortly after the separation of the Moon in his lectures to the workmen at the Goetheanum.* There he spoke of the viscous material of the gradually solidifying Earth, sometimes densifying to the consistency of a horse's hoof and then dissolving again. Thus there was an alternation of densities, and the solidifying of the rocks was not a linear process but a rhythmic progression towards the solid state.

In this alternation something of the life of the Earth as a whole comes to expression. It can be seen that even material such as grains of sand of the fragmented primal rocks was, to a certain extent, subjected to the rhythmic ordering of the life of the Earth. The form which arises under these laws imitates the structure of the old crystalline rocks – namely a granular rock. In the previous epoch, the schist/slate period, these laws had an even stronger influence on the structure of the rock. The schists are often composed of the same substances as granite and gneiss. In these rocks the forces of the surrounding universe – the heavenly bodies – dominate, and out of the soft primal magma call forth, like an enchanted reflection, the crystals in granite and gneiss. In schist the life forces of the Earth itself determine the woodlike grain of the rock. Terrestrial forces

---

* From *Crystals to Crocodiles,* lecture of Sep 23, 1922.

are even more deeply involved in the formation of sandstones, conglomerates and breccias. From the *living* sedimentary forces of the schist/slate period gradually the purely mechanical forces of sedimentation develop.

After the schist/slate period, the life of the Earth itself changed from being plantlike to being more animal-like. This is evinced by the ever-increasing beds of limestone. The forces from the surrounding universe that prevailed at the beginning of the forming of rocks gradually withdrew and the silicious rocks were drawn into the sphere of influence of the purely terrestrial forces. This process can be traced through all its stages. There are sandstones where fractured surfaces of grains have recrystallised as if each single grain was striving to re-form a complete crystal. In the rocks of these sandstones we can find beautiful nests of crystalline quartz which, however, usually only have pointed tips and not the shafts as they have in other rocks. In the Bunter Sandstone we may find cracks lined with these pointed quartz crystals. In the sandstones of the Keuper and Cretaceous formations one has to search hard to find any. In these rocks the power to crystallise diminishes more and more, for the Earth forces of death and gravity begin to dominate.

However, it would be a great mistake if we were to assume that the forces of the heavenly bodies in the surrounding universe had no further influence at all on the forming of rock. What we have described regarding sedimentary rocks does not apply to all other rocks.

To see why, we have to look back at the oldest geological times. Sandstones and conglomerates have a remarkable past. In old strata we can sometimes find a series of rocks which have flattened out boulders embedded in a uniform crystalline mass. The boulders are of the same material as the underlying rock. They are quite obviously part of the primal rock broken loose, formed into boulders and incorporated into the subsequent sedimentary strata. This is a very early form of conglomerate.

Or it can happen that granite or gneiss will be completely fragmented and the constituents – quartz, mica and feldspar – are re-cemented together forming a sort of reconstituted granite. These are called arkoses.

These rocks together with the Rotliegend conglomerates, the Permian Sandstones and the Bunter Sandstone occur at a period of intense volcanic activity and upheaval of mountain ranges. We only know of these upheavals from folds which usually lie hidden in the Earth but can be discovered in mines and sometimes when they are exposed on the surface. The mountains of these ranges have long since been worn down and we find their fragmented remains in the various sedimentary rocks we have described. The rock of these mountains was not as hard as our present rocks. They were still waxy as Rudolf Steiner pointed out. They were mostly broken down by water. The breakdown was not so much a splintering, as a crumbling, for the crystallisation process had only just begun and the various crystals were not yet firmly interlocked. The fragments were suspended in a kind of mire which was quite thick and contained a lot of material in solution. In early times, what solidified out of this mire became dense rock like quartzite for instance, where one cannot see that it consists of minute particles welded together. All gradations exist from quartzite to true sandstones where grains appear as if they had been hard and solid from the beginning.

Such mountain ranges like those before the Bunter Sandstone times arose again later. The younger ranges are well preserved and can be recognised by their sharp and rugged outlines. The rumps of the old mountains are either buried or else rounded and worn like the Erzgebirge, the Bohemian and Bavarian Forests and others. Younger ranges are the Alps, the Caucasus, the Himalayas and, in South America, the Cordilleras and the Andes.

These younger mountains show that the rock masses when they rose up from the depths were still in a condition similar to the original condition of crystalline primal rocks such as granite and gneiss. The younger rocks were obviously still in a very plastic condition since they show beautiful folds in their schists like the older rocks (Plate 1). The strange thing is that some of the younger rocks have been formed in such a way that they have the appearance of very old rocks. In Europe for instance there is a slate from the Black Jura transformed into real mica-schist, while the usual Black Jura slate, as found at the foot of the Swabian Alb, has a foliated crumbling structure. For a long time this transformed slate was mistaken for old crystalline schist until it was discovered to contain the same fossils as the Black Jura.

The transforming of a young formation into the structure of an old rock was explained as due to the enormous pressures exerted during the folding of mountain ranges. It is more likely that the condition of the rock was far more plastic and alive during the forming of these young mountains. The crystallising forces proceeding from the surrounding cosmos could then work upon them, and this is born out by the fact that the largest known rock crystals are found in these younger mountains. If these gigantic crystals had been formed before the folding of the mountain ranges they would have been crushed and ground to unrecognisable fragments.

So we see that even in fairly recent times, in certain parts of the Earth, rock was still not so hardened that it could not respond to these cosmic forces. In the regions of the younger mountain ranges the Earth experienced a certain enlivening which affected all rocks in the neighbourhood.

We cannot here attempt to describe all the true sedimentary rocks. Nevertheless there are still some remarkable phenomena that allow us a glimpse of an earlier condition of what we today call rock. In the province of Minas Gerais in Brazil and in Jhajjar near Delhi in India there is a rock called itacolumite or flexible sandstone. Its appearance is like a slaty sandstone with the pecularity that quite thick slabs can be bent to a certain extent without breaking. Examination under a microscope shows that it is composed of very irregular large and small grains of quartz that are precisely interlocked. Between the interlocking grains lies a thin film of a talc-like material which does not cement the grains but acts like the fluid or film of oil in an accurately fitting joint, and facilitates the movement. In this flexible sandstone diamonds, native gold, specular iron (iron-glance) and magnetic iron ore can be found.

Here nature has left a record of a previous state of matter. There is no doubt that quartz grains were so plastic when the rock was formed that they mutually shaped one another as can be seen under the miscroscope. However, the density was such that they did not mix with the talc-like substance but this remained in the interstices between the soft quartz grains.

Another example shows that even younger rock formations such as the flysch of Alpine foreland regions contain varieties of rock that were not yet hard. Flysch is a conglomerate that originated from a great mountain range that was being largely eroded before the folding that produced the Alps. This rock looks like a coarse concrete of river-worn pebbles tightly cemented together by a calcareous sandy binding material. Flysch consists of an enormous variety of rocks from granite and gneiss to schists, limestones, sandstones, etc. In the calcareous and clayey pebbles one often finds indentations into which neighbouring pebbles fit to perfection (Plate 12). Some of the pebbles had not become as hard as they are today while others were hard enough to cause impressions in the softer material.

In argillacious rocks and marls, which are as variable as the sandstones and conglomerates, we can study all the transitions from true slates to actual mud deposits. In argillacious rocks aluminium silicate, which is the basis of slate, predominates; in marls this is mixed with lime or dolomite. Here also substances are slipping out of the hold of the living form-giving forces and are falling under the dominion of gravity which works purely physically in the depositing of sediment. The structure of these rocks, if one can call them such, tells us clearly what forces have been at work on them. The well-formed crystals in them show that materials were still available in which the forces of the heavens could work. Fine, flaky foliation speaks to us of the vegetative life of the ancient Earth, while dust and debris, covering the most beautiful crystals of gypsum and iron pyrites, point to the death forces of gravity and weight that take over the world of rocks.

10. Volcanism and the Birth of Terrestrial Fire

Volcanism – actively erupting volcanos as we know them today and from historical records – did not exist when rock began to form. As all other phenomena of living and dead nature have passed through stages of development, so also with volcanism.

We have already observed that the concept of the Earth as molten primal magma is no longer tenable, especially when considering the structure of the oldest rocks such as granite, gneiss, gabbro, pegmatite, etc. We have also explained how the precursory stage of rock was gelatinous, malleable or flowing, while heat or warmth was active in it as in a living organism. From this primal condition rocks that are today called igneous underwent a gradual hardening process in rhythmic repetitions. There were times when the soft rock mass densified and other times when it again became quite fluid. This hardening process was caused by a loss of water, air and heat. By re-absorbing water, air and heat, the rock was reliquified.

Such changes, which took place over the whole globe, are to be seen as life processes of the Earth. A characteristic of this more or less living condition of the rock mass was that it took on forms as it flowed, which can strangely resemble the flowing lava forms of active volcanoes today. On the other hand it is evident that these moving masses of granite, greenstones and porphyry never developed into volcanoes, but only partially penetrated the soft masses of the overlying 'rock' and gradually hardened. In this way arose the laccoliths (or concordant plutons), sills and veins so characteristic of these rocks.

A further stage of the development of volcanism was that the soft mass broke through to the surface and spread over wide areas in vast sheets. Such sheets of basalt are in Greenland and extend from there to Iceland and the Western Isles of Britain. Another enormous sheet of basalt occurs is the Deccan Plateau in India. The structure of this rock shows how thick and viscous the original mass had been. In Greenland there are stepped layers of basalt hundreds of metres high. Another phenomenon which also shows the viscosity of the escaping rock flow is that sometimes the mass does not manage to overflow from the crater or fissure but hardens there so that enormous, steep domes of basalt are formed. In these old plateau (or flood) basalts one never finds the quantities of tuff, slag or volcanic bombs that modern volcanoes produce. Slow solidification has produced the huge columns of, for instance, the Giant's Causeway in Northern Ireland, or Fingal's Cave on Staffa off the west coast of Scotland (Plate 13).

We can see from the effect they have had on rocks they penetrated that these flows of rock never had the temperature of a present-day volcano. The neighbouring rocks of these hot streams are modified: they are penetrated, silicified and limestone has become crystalline, but pieces of neighbouring (country) rock caught up and carried along have not been melted. A low temperature is also indicated by the percentage of water contained in all such rocks today. Of course, the country rocks themselves were not yet hardened and were far more subject to the action of heat and water. It was found that when such a flow penetrated a coal seam, changes took place that can be reproduced experimentally at about 500°C. But the actual melting point today of such a rock flow would be 1100°C. Obviously, present-day physical and chemical properties do not make it possible to determine indisputably the real temperature.

With this ancient volcanism we are therefore dealing with a

state of matter which no longer exists today, or perhaps only at a certain depth in the Earth. The essential thing is that there was a much more intimate union of mineral matter with water, gases and heat in the flowing rock. The union of the four elements was due to the much greater 'life' of the whole Earth. The flowing of the soft rock was a life process of the Earth organism.

Organised interplay of the four elements is a characteristic of every warm-blooded creature. In the higher animals and human beings the mineral parts (bones and salts) water and gases are maintained together with warmth in a living union. None of these four elements alone can constitute life.

There are present-day volcanic phenomena which show that the magma which bursts out of volcanoes is in a condition that can only be approximately reproduced artificially in the laboratory. Measurements taken in 1910 under the Canadian geophysicist R. A. Daly in the lava lake of Kilauea on Hawii gave the surface temperature of the boiling lava as about 1200°C, but at about eight to ten metres below the surface of the molten lava it was about 100°C less. The highest temperatures (1300 – 1350°C) were in the four-metre high gas flames above the surface of the lake. The burning gases contained about 60% water, 10% carbon monoxide, and 3% oxygen, the remainder being nitrogen and carbon dioxide. The average temperature of the molten lava in the interior of the lake was about 1050°C. Further measurements in an African volcano yielded similar results.

The individual ingredients of such a lava consist of olivine (a magnesium silicate), calcium feldspar, hornblende and other minerals. Their individual melting points lie between 1900° and 1200°C, but through the high water and gas content the melting point of the lava is appreciably lowered. There is such an intimate interpenetration of the four elements that we must realise that this lava is in a very special condition. The latent heat becomes *free* when the lava comes into contact with the atmosphere. The original condition decomposes, releasing heat, steam and gases. Another significant point is that the molten lava contains many ferrous compounds which on contact with the air are transformed into ferric compounds thus creating more heat. This is the cause of the afterglow of lava streams that can be seen for days or even a week after an eruption.

In the laboratory such molten rock with its content of water and gases needs to be heated in a closed vessel with an enormous pressure building up. Under such pressure the rock actually melts at about 1050°C. If, however, in nature such a pressure existed in a lava lake, the molten lava would be hurled up some hundreds of metres into the air. But the lake remains comparatively calm with jets rising to only a few metres from which the wind draws out the lava into glassy threads. The native Hawiians call these Pelé's hair after the goddess of volcanoes.

Such simmering lava lakes have been observed at times in Vesuvius and other volcanoes. The heat produced is unimaginable. It has been estimated that the lava lake Halemaumau on Hawaii produces 300 million calories per second. Where does this heat come from?

The second phenomenon we would like to mention is the eruption of a heavy emulsified mixture of red-hot rocks and gases with no molten lava, called a pyroclastic flow. This mixture can be so heavy that it can rush down the mountain side like a lava flood, destroying everything. It was such a fiery eruption which destroyed the town of Saint-Pierre, Martinique, with its 26 000 inhabitants in 1902. An impeding lava plug in the crater of the volcano Mount Pelée caused a lateral explosion and a black cloud of ash and stones, flashing lightning, burst out and sped down the mountain side. With a speed of 150 metres per second (335 mph) this hurricane cloud leaped on the town 6.5 km (4 miles) distant which was completely destroyed in a few seconds. The effect of the heat indicated a temperature of about 800°C. The neck of a wine bottle was bent like a heated candle, iron tanks were shot through by stones. Wooden planks were driven through tree-trunks. The only survivors were a prisoner in an underground cell and a severely burnt man on the edge of the city.

Such 'clouds' of glowing materials show a condition which contradicts all physical laws. Indeed they are a sort of 'emulsion' of solids, gases and heat. This mixture behaves sometimes like a liquid, sometimes like gas, at other times like a solid body. Nevertheless it is a unity and it even has its own thunderstorm centre. The lightning phenomena familiar in other eruptions show that there are peculiar atmospheric conditions in the vicinity of a volcano at the time of an eruption.

Rudolf Steiner made an important observation regarding the causes of volcanic eruptions. In two lectures to the workmen at the Goetheanum he described volcanic activity as dependent on forces which act from the solar system, due to the Sun and certain angular relationships of the celestial bodies.* He illustrates this by the well-known phenomenon of the solfataras north of Vesuvius which begin to steam when a piece of paper is lit near them. The warm air from the burning paper produces a slight reduction of atmospheric pressure over the surface of the ground so that the gas begins to emerge more strongly. A local reduction of pressure, brought about by astronomical relationships, is the real *cause* of an eruption. Because of this it also happens that a local thunderstorm and downpour of rain occurs over the erupting volcano even if the sky is cloudless. In the second of these lectures Rudolf Steiner came out against the theory that the interior of the Earth is molten, and pointed out its incompatibility with the specific weight of the Earth. The Earth is too heavy to be molten in its core.

According to this, one can picture that concentrations of heat arise which, after a long incubation, are released by a particular relationship of celestial bodies. These fire processes arise locally at particular spots. Such an explanation of course contradicts the concepts and theories of modern science which make of every fiery crater a blast furnace heated from below.

* *From Mammoths to Mediums,* lecture of June 2, 1923, and *From Sunspots to Strawberries,* lecture of Sep 18, 1924.

One is really dealing with an organic conception of this natural phenomenon. It sees the Earth inter-related with the surrounding world. In this sense one can consider volcanism in its present form as a life process of the Earth. The *old* volcanism was quite different. It was also an expression of the life of the Earth, but it did not have the fiery character of today.

The awakening of the terrestrial forces of fire took place very late. What we have described at the beginning of the chapter applies particularly to the younger igneous rocks such as trachyte and basalt. They are transition stages (trachyte) which in the beginning were still very much like granitic rocks. The trachytes are light coloured and rough textured rocks composed of feldspar, hornblende and some magnesium mica. The Siebengebirge on the Rhine is almost entirely trachyte. The compounds of this rock often hold sufficient amounts of water to show that it was never in a molten state. Even though trachyte is not so widely distributed as basalt, yet there are numerous forms and transitions which connect it on the one hand with the younger porphyries and on the other hand with the basalts which follow it. There is also the well-known phonolite or clinkstone which is so dense that it has a metallic ring when struck.

In the incredible variety of these young igneous rocks (which with the basalts and present-day lavas becomes countless) there is a phenomenon comparable to that of the lamprophyres, etc., which we have spoken of in connection with the metamorphoses of granite. To describe all these varieties would exceed the scope of this book. We can only select those which have a particular bearing on our subject.

The pale siliceous and iron-deficient trachyte represents a transformed repetition of the granitic or older granular rocks.

What is typical of basalt is the much lower silica content and the generally high iron content. Because of this, nearly all basalts are dark or even black. Their mineral content, which in the main consists of augite, calcium feldspar (labradorite) and magnetic iron ore, brings these rocks into a close relationship with the greenstones (gabbro, diabase, serpentine, etc.). This is shown in a very striking manner. We remember that the greenstones are associated with the plantlike stage of the Earth life. We find in them many mineral forms reminiscent of wood, for instance in the fibrous asbestos. In basalt this fibrous structure is magnified to gigantic dimensions and produces basalt columns. The five to eight sided pillars which can be up to 100 metres (300 ft) high have nothing to do with crystal formation though they are forms born of forces acting on the Earth from outside. Even as plants are in a certain sense drawn by their stalks out of the Earth by the power of the Sun, so likewise are the thousands of stemlike pillars of the basalt formations drawn out by the power of the Sun. Imaginatively one can think of the basalt pillars as the saurians of the rock world. They appear mainly at the time of the saurians (Cretaceous and beginning of the Tertiary).

Rudolf Steiner described the formative process of basalt as a freeing of the Earth from excessive Moon forces. The powers to effect this freeing proceeded from the Sun and entered into the rocks which were formed at the time when the Sun had its strongest influence on the Earth. These rocks appeared when the Moon was still part of the Earth and thus have become the densest and toughest of rocks and contain the hardest and most fire-resistant minerals and metals. In the partial transformation of the old rocks into basalt – and especially columnar basalt – by the Sun, lies the secret of the freeing of the Earth from 'excessive Moon forces'.

Following basalt, in the course of the Tertiary up to the Ice Age, the lava of present-day active volcanoes gradually developed. In this process the original close 'organic' union of the elements has fallen apart. Terrestrial fire was thus born out of the world of rocks just as the great water masses were born out of the misty atmosphere of Atlantis, and descending, engulfed that continent and caused the tremendous cooling of the Ice Age.

In the lavas one can recognise certain basic types. They are primarily either trachitic or basaltic and thus show that the lava process developed out of these previous processes. But in actual fact there are as many different lavas as there are active volcanoes. This is due to the fact that now the terrestrial fire really takes hold of *all* rocks and transforms them. The result of these transformations are slaggy, porous materials such as pumice, innumerable forms of lava from smooth or ropy (pahoehoe) to block lava, ashes, scoria and volcanic bombs. Nearly all these materials are more or less porous and cellular, sometimes grotesquely so. The innumerable cavities in lava are due to the escape of gas from the viscous, molten mass immediately after its ejection or outpouring from the crater. The spongy structure of the solidified lava shows that the viscous mass was full of gas, and, on the hardening of the flowing lava, rock appeared whose cavities are distinctly reminiscent of those in melaphyre and some porphyries with their beautiful agates and embedded minerals. These latter structures, reminiscent of organic forms created by the living Earth with the processes of heat, air and water, can no longer be formed in lava; terrestrial fire has become the enemy of life.

11. Natural Radioactivity and the Process of Germination

With sedimentary rocks we have been studying a stage of development that we can only describe as one of breakdown and disintegration. This process which we can still observe daily in the mountains, and hourly in the lowlands in our streams and rivers, embraces the second half of the total history of our planet. We have tried to sketch in broad outlines the first half of this history as the building up of the past life-stages of the Earth. This process ceased some tens of thousands of years ago and we find ourselves now in the age of the Earth which corresponds to the years just past the middle of a human life, around the 35th year. From this time on, depletion and decomposition begin increasingly to take effect. This, however, is necessary to achieve full human maturity. What has not been built up physically and spiritually until then can no longer be done later. To mature rightly we have to transform the physical growth forces of youth that begin to wane at this age into spiritual qualities.

Just as human beings have to consciously take charge of their lives from the middle of life onwards, so they must learn consciously to counter the dying forces of the Earth in order to preserve it in a habitable condition for some time still. We need not fear an ice or heat death for the ageing Earth, for the future of the Earth runs parallel to human development. When finally the Earth reaches whatever end is in store for it, then humans as spiritual beings will have freed themselves from their present earthly needs and will be able to pass on to a different state of existence. But at present we think very little about countering the disintegrating forces of the Earth, even making increased use of these forces. What for thousands of years was done slowly only by the weathering of the rocks, we now carry on by mining, deforestation, altering of water courses and unnatural methods of plant and animal husbandry. 'The rape of the Earth' is what the German sociologist and economist Alfred Weber once called our present world. Humankind took the last, dangerous step towards the disintegration of the Earth by making use of the radioactive powers of disintegration of matter.

It is not a criticism of the achievements of science and technology that they act destructively on the life of the Earth. Such advance is necessary because from the negative results of this progress we ourselves must learn to act positively. In order to do this we must ask what really is natural radioactivity.

There can of course be the most diverse answers. We shall attempt to find one that can be read from the natural phenomena themselves, without reference to any existing theories on the nature of radioactivity. Naturally there are other possibilities, but we shall leave these aside for the present.

Among the many chemical elements there are a few which are known to give off various radiations. These invisible radiations produce either electrical phenomena (beta and gamma rays) or are perceptible through light effects caused in certain other substances (alpha particles). The principle elements which give off these radiations are uranium and thorium. Also for example, the widespread element potassium (so necessary for life) and the rare elements rubidium and samarium have radioactive properties, though very much less than uranium and thorium.

The radiations of these elements are accompanied by the breakdown of the original element which is transformed into a series of radioactive substances of varying length of life and (in the case of uranium and thorium) ends up as lead. This end product of the breakdown is no longer radioactive. A further end product of uranium and thorium breakdown is the inert gas helium which likewise is no longer active. The radioactivity of these elements – that is, their continuing breakdown – cannot be influenced by any process of chemistry or physics. A piece of uranium can be heated to melting point, or cooled to –200°C, or subject it to a pressure of 1000 atmospheres – its breakdown is neither hastened nor retarded, nor can it be halted by any other means.

These strongly radioactive elements, uranium and thorium, are native to the oldest primal rocks of the Earth – the granites and pegmatites. In these rocks they are extremely finely distributed and only in relatively few places are they sufficiently concentrated to be quarried or mined. Through the mechanical breakdown of rocks, before they had really hardened, uranium and thorium found their way into younger rocks such as the famous carnotite sandstones of Colorado. However, the main sources of uranium and thorium for the production of atomic energy are to be found almost exclusively in granites and pegmatites.

Today over a hundred minerals are known to contain uranium, and about fifty containing thorium. Original radioactive minerals (rather than decay products from them) show something that is only found in radioactive minerals. These minerals, like uraninite (formerly called pitchblende), thorianite, thorite, cleveite, whether embedded in rock or freestanding crystals, appear to be fully formed on the outside, but within have the appearance of solidified jelly-like or pitchy substance (hence the name pitchblende). If, using X-rays, we make a diagram of the lattice of these crystals, it transpires that in their interior there is no crystalline structure. In mineralogy these crystals are termed isotropic. If such isotropic crystals are heated a little, they glow, and the lattice, which according to the outer form should be there, now appears under X-ray examination.

We have the remarkable fact that these minerals, which once upon a time were crystalline, through their radioactive decomposition became jelly-like or colloidal. In fact these radioactive minerals, like all minerals, originally crystallised out of a gelatinous condition. This can be recognised from the outer form of the mineral. Radioactivity reversed the process and they reverted inside to an amorphous gel. The fact that such isotropic minerals when heated re-establish a crystalline structure is further proof that the primal molten rock theory is untenable.

The occurrence of radioactive minerals in the primal rocks and this isotropism which we have just described, are, in our opinion, important basic phenomena which lead towards the understanding of one aspect of radioactivity.

In order to do so however we must turn aside to an apparently unrelated subject, to the process of germination in the plant. By looking at present-day life-processes we recapture something of the living Earth of the past.

Seed formation is an extraordinarily involved process, part of which resembles a sort of mineralisation. When the seed in the ovary has reached the stage when the whole of it is still green and the ripening process is just about to begin, it is called 'milky'. The first garden peas we enjoy in the springtime are in this condition. The unripe milky seeds are not viable if they are harvested and dried, for they lack a vital process which only takes place during ripening. In the milky seed the seedling is fully formed, but the nutritive substances which accompany it in the form of starch, protein and fat are still in the colloidal or jelly-like state. On examination of such a seed it is easy to determine that it contains as yet very little mineral matter. During the subsequent ripening there is not only the obvious drying of the seed, but starch and protein begin to form and gradually harden out of the jelly-like colloidal state. Starch forms typical starch granules, with their fine inner layers and round or many-sided shapes. Protein congeals into real crystals which are known as crystalloids. (For example the aleurone granules of cereals whose nutritive value is so important.)

At the same time as this drying and ripening, starch and protein bodies become permeated with certain minerals such as calcium, magnesia, potash, phosphoric acid, silica, etc. One can thus characterise ripening as a sort of mineralising of the seed. Not only its substance but its whole inner structure becomes rock-like or 'earthy'. This is necessary because the seedling which remains unaffected by the drying process will later, when it germinates, need this 'earthiness' of the seed. Every seed, in ripening, builds for itself its own little Earth which supports the early life of the seedling until it finds its union with the greater Earth.

A ripe seed sown in spring begins at first to swell under the influence of warmth and moisture and, after a few days, the first thing to appear is the radicle. Before the radicle pushes through, however, something important has taken place in the nutritive substance of the seed. Starch and protein have changed their structured mineralised condition and reverted to a jelly-like state.

The new jelly-like state of the germinating seed bears a great resemblance to the milky state, but this time is not followed by ripening. During germination the nutritive substance of the seed disintegrates into water, carbon dioxide, ammonia and salts. We will not discuss here how the intermediate stages of this breakdown affect the seedling.

This gradual process of disintegration of nutritive substance and the associated root growth is accompanied by something that is not outwardly visible. The germinating seed sends out radiations. The existence of these germination and growth radiations was discovered by biological and physical tests. The first to describe this accurately was the Russian scientist Alexander Gurwitsch. These radiations which he called biophotons are closely related to ultra-violet rays in that they can pass through quartz crystals but not through ordinary glass. They have a favourable effect on the growth of other living organisms and plant organs. It is probable that the traditional custom of attaching a grain of corn to a cutting that is to take root is not only connected with what are known as growth substances (auxins) but also with these biophotons.

Thus we have, in the processes taking place between the milky state and germination, a true reflection of these other processes that we encounter within the mineral world in natural radioactive compounds.

If we remember what was said in Chapter 3 regarding the life processes that led to the formation of granite and kindred rocks, and then consider this ripening and germination of seeds, a significant picture emerges. Following Rudolf Steiner's suggestion we tried to show that the primal rocks originated out of the tremendous, though at that time undifferentiated, flowering stage of the whole Earth. From this viewpoint we might say that the nature of the old crystalline rocks has something in common with a seed. If these rocks constitute something like 90% of the mass of the Earth, we can see the Earth as a totality as a giant seed. This 'seed' Earth ripened during the time which elapsed between the first gel condition of the rocks and their present solid state. This hardening of the rocks is a very gradual process and at different times and in different places alternates with softer conditions. There is evidence that at the times of the great mountain range upheavals the rock mass in certain areas was still in (or had reverted to) a pliable condition, for otherwise the crystals and minerals in the hollow cavities would have been completely crushed. Further evidence of later hardening of the crystalline rocks is to be found in the temples of earlier civilisations where, for example, in Egypt and Peru, stones were accurately dressed from basalt, diorite or granite without iron tools. These buildings date from between 3000 BC and AD 1000.

Hardening, together with mineral formation, continued right into the Christian era but the processes of disintegration, weathering and mechanical break-down of rocks began much earlier. It began from the time when no further new rocks were formed as a result of *living* processes.

The minerals which today are radioactive were present in the primal rocks during this whole process of hardening. They are evidently a sort of ferment which, extremely finely distributed, works towards the dissolution of the Earth as a seed. One might say that the Earth-seed is beginning to germinate so that out of this dissolution a new future existence may arise. We need not discuss the nature of this existence here: there is much about it in Rudolf Steiner's works.

From Rudolf Steiner's investigations it appears that before our present epoch there were two other great cultural epochs which have disappeared without trace. In the older of these, the Lemurian epoch, the conditions of existence were completely different. Human beings still had the capacity to influence the form of an animal. They had power over the reproductive forces of animals. The Lemurian epoch disappeared through the misuse of these powers. In the succeeding Atlantean epoch, on a continent between Europe and America, humankind had the power to work creatively on the plant world, being able to control the germinating and growth forces of plants which were used for technical purposes. The Atlantean continent and its civilisation perished through the misuse of these forces between the eleventh and eighth millenia BC. We have to thank these two prehistoric epochs for our domestic animals and food plants.

Our own culture is based entirely on the mastery of the mineral kingdom. Now, however, after the discovery of atomic energy, which is based on the natural radioactivity of certain minerals, our epoch begins to make use of the equivalent in the minerals of the 'powers of germination.' When these forces become misused then our civilisation, too, will perish. By the time this happens humankind must have reached new levels of existence. Disintegration, destruction and death are the foundations of new life.

12. Meteoritic Phenomena

Life in our present-day Earth is no longer active in the world of rocks but only in plants and animals and in a certain way in air and water, but there is one way in which it still enters into the world of rocks and metals – the appearance of meteors in the atmosphere.

Before we go into the nature of this process we must be quite clear about what is meant. Meteoric phenomena are usually understood as comprising all 'falling stars', that is, shooting stars and fire balls but not comets whose behaviour is altogether different.

A shower of meteors or fireballs ray out from a particular region of the stars, the radiant. Astronomical research on meteors, especially the work of the German astronomer Cuno Hoffmeister, showed that there is one stellar region from which nearly all meteors come that reach the Earth (when they are called meteorites). This region lies in the constellation of Scorpio.

When, around August 12 of each year, we see the shooting star showers of the Perseids, we must understand clearly that these are light phenomena from which no meteorites fall to the Earth. The same holds good for almost all shooting stars during the course of the year.

The fall of meteorites (stone or iron) is always accompanied by loud noises (explosive thunder, rushing and hissing) and intense light phenomena and, in the case of fireballs, the forming of a tail. The thunder clap which accompanies the extinction of light of the meteorite is not due to an explosion of the meteorite, but rather to the *implosion* of a luminous space. What later forms the metallic or stone meteorite enters the atmosphere as light (physicists would say in a highly ionised state or plasma) and in the moment of implosion suddenly condenses as solid matter. An observation of the Treysa meteorite near Kassel on April 3, 1916, recorded a fireball of 1000 metres (3300 ft) diameter at a height of 50 km (30 miles) which rapidly diminished to 400 metres (1300 ft) and at a height of 16 km (10 miles) imploded, and dropped as an iron meteorite of 63 kg with a diameter of 36 cm (139 lb, 14 in). The shower of meteoric stones at Pultusk in Poland in 1868, which scattered some hundred thousand stones, originated from a light phenomenon 300 metres (1000 ft) in diameter at a height of 50 km (30 miles).

Thus it appears that a huge body of light suddenly condenses into the solid state, rather as in a case of sublimation (the transformation of a gaseous substance into solid crystals). The impression of this sudden solidification is to be seen in the structure of meteoric iron and meteoric stones. In the case of meteoric iron there is an elaborate complex of large crystals in the interior of the meteorites (Plate 11) built up of varying proportions of iron and nickel. This Widmannstätten structure cannot be produced artificially. In fact if one heats a piece of meteoric iron showing Widmannstätten structure this will be completely destroyed at about 900°C and what remains will be indistinguishable from ordinary terrestrial iron. This tends to prove that meteoric iron was *not* molten before it hardened. It originated from quite a different condition of substance than is known to us on Earth.

In the case of meteoritic stone the sudden condensation is suggested by the unique texture of the chondrites. These chondrites contain small globular bodies called chondrules ranging from microscopic to pea size, which often have an inner fibrous radiating structure. They are embedded in a ground mass of the same composition. Other chondrites are not crystalline in structure but resemble tuff in that they are porous. The phenomenon of chondrites is entirely confined to meteoric stones and does not occur in any terrestrial rocks.

Such chondrites are really a sort of emulsion of finely

distributed iron and other minerals which can only have been formed in a non-gravitational field. The heavier and lighter constituents did not separate according to the law of gravity, but remained mixed like an emulsion of water and oil. By contrast, although basically every granite, syenite or diorite is made up of heavy and light minerals, in these rocks one never finds chondrules which are so characteristic of the meteoric rocks. And in these terrestrial rocks only intergrown crystals of various minerals are found. It is the forming of chondrules which gives the clue to the rapid condensation into droplike forms.

With meteors this whole process only reaches the stage of luminosity, but the showers of meteorites from the constellation of Scorpio reach the density of matter. This does not mean that no matter ever reaches the Earth from shooting stars. But their individual mass calculated from their light and velocity amounts to no more than a fraction of a gram. It may be rather rash to draw conclusions regarding 'mass' from pure phenomena of light and velocity in space, but extremely finely distributed meteoric dust exists in the Earth's atmosphere.

Meteors or shooting stars and meteorite showers differ in other respects. Shooting stars have a lower velocity – average 41.6 km per second – and an almost parabolic path, while the large meteorites reach the Earth at over 62 km per second following a hyperbolic path. Most of the meteor showers have an indisputable connection with disintegrated comets, while it has never been observed that the origin of a meteorite shower has any connection with comets.

The maximum annual fall of meteorites occurs in June when the Sun is in the constellation of Taurus. This is a minimum period for meteors or shooting stars. The number of meteorites begins to increase in April and then in July decreases to below the April figure. It is important to realise that in June the Sun appears in Taurus, and the meteorites originate from the opposite direction, from Scorpio, falling as meteoric iron and meteoric stone.

This does not mean that such showers only occur from April to June, for there is a certain degree of activity the whole year round from the direction of Scorpio but it only reaches its maximum when the Sun is in Taurus. These variations in the course of the year suggest a living organic process that is no more precisely calculable than the growth of a tree.

To complete the picture it should be noted that the area of the heavens lying between the constellations of Scorpio and Taurus is notable for starless areas and dark nebulae. On the basis of observation, modern science has come to the conclusion that there must be a greater density of matter in these areas obscuring the stars behind.

What do we know about the origin of meteorites? Let us begin with what can be learned from the composition of meteoric stone which is more common than meteoric iron. Most meteorites are chondrites which, as we know, are a mixture of finely mingled wisps of iron and rock.

The minerals of which this rock is composed are, strangely enough, always the same. What sort of stones are these which 'fall from heaven'? They are usually diabase – a rock of the family of greenstones which we dealt with in Chapter 3. In this family of greenstones the older 'relatives' of diabase are gabbro and serpentine, the younger relative is basalt. Similarly some meteorites resemble gabbro, others basalt. They also contain a characteristic suite of supplementary minerals. Diamonds occur very occasionally, graphite more often, also ferro nickel, iron sulphide (magnetic pyrrhotite) which here occurs as troilite, magnetic iron ore and iron chromate (chromite). All these minerals also occur on the Earth in the various gabbros, diabases and basalts.

However, meteoric stone and some meteoric iron also contain minerals which are unknown among terrestrial rocks. There are, for instance, oldhamite, daubréelite, schreibersite (or rhabdite). The principle minerals which constitute the rock of meteorites are plagioclase (calcium sodium feldspar), enstatite, bronzite and hypersthene (all magnesium iron silicates), diopside (magnesium calcium silicate), augite (magnesium iron aluminium calcium silicate), olivine (magnesium iron silicate), fosterite (magnesium silicate). These are the same minerals which are also the constituents of the terrestrial gabbros, diabases and basalts. Olivine plays a particularly prominent part in meteoric stones. It was found in gemstone quality (chrysolite) in the meteorite of Krasnoyarsk in Siberia.

The remarkable thing is that ferro-nickel is found in terrestrial rocks such as dunite and peridotite which consist mainly of olivine. Examples of terrestrial ferro-nickel are awaruite of New Zealand, josephinite of Oregon, and souesite of British Columbia. Then there is the ferro-nickel found in the basalts of the Island of Disko (Western Greenland) and of the Habichtswald near Kassel, Germany.

If we bring all these facts together it appears that in the meteoric stones there is the equivalent of the greenstones lying deep within the Earth between granite, gneiss and the crystalline schists.

Let us remember that the greenstones by their very nature and structure are closely connected with the tree stage of the Earth. We also pointed out that the organic woodlike structure and the composition of these minerals – greenish magnesium iron silicates – are the characteristics of the plant stage of the Earth. We mentioned that former life stages of the Earth were recapitulated; in this case, the stage when the Earth, Sun and Moon were still one heavenly body. We then described how this stage was followed by the formation of schists and slates and that after the separation of the Moon-Earth from the Sun, there arose the Coal Measures. The formation of greenstones however falls entirely in the time before the separation of the Moon-Earth from the Sun and before the formation of the Coal Measures. Basalts are later transformations of greenstones as we described in the chapter on volcanism.

We can say therefore that the origin of greenstones is

connected with the Sun evolution of the Earth and their diversity comes about through the threefold recapitulation of former life stages.

Rudolf Steiner stated that the substance of meteorites and shooting stars was radiated out into space from the Sun. This radiation should be understood as a light process – substance in a state of light. The modern physicist would speak of a highly ionised state of matter assumed to be in space. The radiation of the light substance from the Sun is connected with sunspots. One might take it as a reaction to the formation of sunspots. The light substance which is thrown out from the Sun appears within our solar system as meteors and shooting stars.

Rudolf Steiner acquired this information by spiritual research, and first spoke of it in 1923. It was corroborated in the early 1940s by the work of Bengt Edlén, Bengt Strömgren and Max Waldmeier who, through spectrum analysis, found that the light emanating from the corona and the edges of sunspots showed both qualitatively and quantitatively the same elements as are present in meteors.

Thus it appears that today on the Sun, substances are being formed in the same way as during a past life stage of the Earth when the two bodies were one. Through the living interaction of Sun and Earth the family of greenstones came into being. What in those days was achieved gradually through long periods of time by the interaction of Earth and Sun, is today a process proceeding from the Sun and affecting the life of Earth and human beings.

The formation of meteoric stones is a metamorphosis of that past process which produced a particular family of rocks on Earth. How does it happen then that meteorites are mainly formed in and proceed from the direction of Scorpio? Let us follow a clue given by Rudolf Steiner when he described how the great periods of development stand under the 'rule' of particular areas of the zodiac. For Old Saturn it was Leo; for Old Sun,

Scorpio (in earlier times called the Eagle); for the Old Moon, Aquarius, and for the Earth, Taurus. One should picture these areas as the 'seat' of those creative spiritual forces guiding each particular stage of development. So the Sun period was under the sign of Scorpio and the present Earth period under Taurus.

Thus it is from Scorpio that the forces are active which formerly ruled the joint life of Sun and Earth. The Sun-life of the Earth proceeded from Scorpio and what was expelled from this life as substance condensing right down to the mineral state, received its impulse from this region. What today radiates from the sunspots as light-substance is received by the constellation of Scorpio and reflected to the Earth from there. Taurus, the Earth sign, lies exactly opposite Scorpio. On this Scorpio-Taurus line, right across the solar system, is the field of starless areas and dark nebulae which we mentioned above. We can only understand this phenomenon if we consider that there are qualitative differences in the different directions within the zodiac. It is along the Scorpio-Taurus line the densification takes place of the solar light-substance into earthly rock and metal.

Astromomers know that the meteor showers from Scorpio are the only ones originating in interstellar space having *no* connection with comets, but cannot explain why. They know that meteoric stones always consist of greenstone materials and never of granite, gneiss, schist or limestone, but again they cannot explain why. If one brings together the findings of natural science and spiritual research an explanation begins to emerge.

13. The Mineral Composition of Rocks

About 95% of the known rocks belong to the igneous group. It is striking that the foundation of our world consists of a particular class of rocks characterised by a crystalline, granular structure. The quantity of slates and shales, most of which in their basic composition resemble igneous rocks, is small – a mere 3.7%. Likewise the sandstones, containing an average of up to 80% silica, constitute only about 0.7% of the total rock. What remains are the limestones – 0.2% consisting of about 80% calcium carbonate.

We will take these four main groups of rocks together and see of what minerals they are composed. Initially we will concern ourselves not with the constitution of the individual rocks but with the minerals which are the main building materials of the world of rocks as a whole.

The figures which follow are the outcome of many decades of research. They summarise the results of hundreds and thousands of analyses by methods able to detect fractions of milligrams of substances present in the sample.

One of the most important findings is that feldspars account for well over half (59.5%) of our rocks. The next largest group comprising augite, hornblendes, olivine and mica amounts to 20.6%. After these minerals, which are all compounds of silica, come pure quartz (also found as rock crystal), sandstones and loose sand, with 12.6%. These three groups of silicic minerals together form 92.7% of the total rock mass.

The remainder (about 7%) is apportioned between approximately 4% iron ores, 1.6% calcite and dolomite, 1% clay (loess, loam, potters clay, kaolin, etc.), accompanied by widely distributed minerals such as garnet, apatite, zircon, titanium ores and manganese ores which make up the remainder. These figures appear at first very abstract and meaningless, but if we consider them in relation to past stages of the life of the Earth they begin to speak.

What does it signify that nearly 60% of rock consists of feldspar? In order to relate this to our earlier studies of the mineral-plant and the plant-animal life of the Earth we must take into account the composition of this mineral. Feldspar and its relatives, including the feldspathoids, fall into three different basic groups. These are potassium feldspar – orthoclase (a potassium aluminium silicate), sodium feldspar – albite (a sodium aluminium silicate) and calcium feldspar – anorthite (a calcium aluminium silicate). These aluminium silicates do not occur pure in nature, for all feldspars contain two of these other components in greater or smaller proportions. These two components are either potassium and sodium, or sodium and calcium. Thus potassium feldspar (orthoclase) in practice is a potassium-sodium feldspar with a little sodium, and sodium feldspar (albite) always contains a certain quantity of calcium. Similarly, calcium feldspar (anorthite) always contains a little sodium. All this goes to show that in nature there are no pure substances as we, with our love of classification, would like to imagine.

Between orthoclase and albite we have sanidine and the perthite feldspars, and between albite and anorthite the plagioclase series with oligoclase, andesine, labradorite and bytownite. Besides these there are the feldspathoids such as leucite, nepheline, sodalite, nosean and hauyne. Leucite is really a potassium feldspar with less silicon dioxide, and nepheline is a sodium feldspar also poor in silica; sodalite, nosean and hauyne are nephelines containing respectively sodium chloride, sodium sulphate and calcium sulphate.

One should not be put off by all these names which are basically unimportant, they are mentioned for the benefit of those who would like to pursue these matters in greater detail.

It can be seen that the above series of feldspars, as described, is an expression of the development of rocks in accordance with their underlying life processes. The path from potassium feldspar to sodium feldspar and on to calcium feldspar is the same as the progression from oldest granites to greenstones and porphyries, and on to trachyte, basalt and lava. Here in the minerals we see the same progression as in the rocks – an increase in calcium and a decrease in silica. This represents the transition in the underlying life processes from the mineral-plant to the plant-animal and finally to the more or less pure animal life processes. Remembering that it is the feldspars that owe their origin to the old flowering stage of the Earth's life, it is clear that the plant nature, which expresses itself most purely in potassium feldspar, is gradually displaced by an animalising tendency expressed by the sodium and calcium-rich feldspars.

Thus we find potassium feldspar (orthoclase) mainly in granite, gneiss, syenite, some porphyries and trachyte, but sodium-calcium feldspar (plagioclase) in its various modifications in greenstones, crystalline schists, porphyries and basalts. The strange feldspathoids – leucite, nepheline and the sodalites – are to be found mainly in young volcanic rocks and in recent lavas.

Such a rough classification should not be regarded as a rigid system; it is only a broad outline which, through the variability of living things, is continually modified. It is just these exceptions to the rule that are the surest sign that the rock-forming processes did not originally follow *chemical* laws, but laws of *life* which were different from those we know today. If only chemical laws held sway, the world of rocks would present a much greater uniformity and simplicity.

In connection with the feldspars it should be noted that their inner crystalline and molecular structure, from the point of view of mineral chemistry, has a peculiarity first described by the

Russian geochemist Vladimir Vernadsky. The inner structure of feldspar is based on a ring (the kaolin nucleus) consisting of aluminium oxide and silicon dioxide. This substance occurs naturally as kaolin or china clay. It is also the basis of all plastic clays, potters clay, brick clay, etc. It is significant that this plastic substance has an inner structure which is not found elsewhere in the inorganic world. We find a ring structure primarily in proteins – the substance that is the foundation of all life, and which is the most plastic substance known. This phenomenon indicates that the feldspars have retained something of the life out of which they emerged.

When we pass on to the second large group of minerals – mica, augite, hornblende and olivine – we find materials and structures with totally different forms. While in the feldspars aluminium oxide plays a major role, it loses all significance in this second group of minerals. In its place in the main two substances appear – magnesium and iron. The transition between the first and second groups is formed by mica. In it there is still the kaolin nucleus and thus the important substance aluminium oxide. A typical potassium mica (potassium-aluminium silicate) is the light-coloured, silvery muscovite which is practically free of iron. It is found exclusively in granite, pegmatite, gneiss, mica schists and phyllite and is often accompanied by potassium feldspar (orthoclase). Muscovite is, however, entirely absent from the greenstones and porphyry group and the greenstone slates. In some pegmatites the mica sheets are as much as a metre across but they do not reach the tremendous size of the feldspar crystals in these rocks. (We mentioned a whole quarry in the Urals working in a single gigantic crystal). Sodium mica (paragonite) and calcium mica (margarite) play only a local and subordinate role, likewise lithium mica (lepidolite) and lithium iron mica (zinnwaldite). This latter however is a transition to the coloured magnesium iron micas, biotite and phlogopite. Phlogopite is the mica which forms probably the largest crystals, for instance at Sydenham, Ontario, where crystals of 1.5 × 2.5 × 5 metres (5 × 8 ×16 ft) occur, giving mica sheets of 1.5 × 2.5 metres. The real magnesium mica (biotite) is a major or minor ingredient of many granitic rocks, some gneisses and mica schists as well as mica syenites and diorites, porphyries and trachytes. The colours of these magnesium iron micas are always dark – brown, green or black.

Related to magnesium mica are the chlorites – fine green flaky minerals bearing many names. The thin green film observable on many rock crystals and feldspars in the Alps is due to chlorite. Other minerals of somewhat mica-like structure are talc (soapstone or steatite – a porous magnesium silicate coloured green by iron) and serpentine in the form of foliated serpentine (antigorite).

There is another form of serpentine which makes a bridge to the pyroxenes (augites, etc.) and amphiboles (hornblendes, etc.). This is the fibrous crysotile-serpentine whose green woodlike formation in dense serpentine is a matrix for asbestos – rock wood and mountain leather. Here we arrive at the significant point where the plant-forming forces of the whole living Earth reveal themselves most clearly in the mineral kingdom. In serpentine asbestos, as in hornblende asbestos, the crystal-forming forces of the surrounding cosmos are almost completely neutralised by the formative forces of the life of the Earth itself.

The groups of the pyroxenes and amphiboles are divided into magnesium iron silicates, calcium iron silicates, calcium magnesium silicates and sodium silicates. Most of these minerals are of a lighter or darker green colour, or even almost black; some show a brownish tinge. Many tend to have fibrous scaly cleavage surfaces giving them a satiny iridescence like mother of pearl. One can have the impression that these minerals, in the same way as chrysotile and asbestos, were moulded by organic formative forces and not much influenced by crystal-forming forces.

The magnesium and magnesium iron silicates – enstatite,

bronzite, hypersthene – are typical ingredients of many gabbros, norites, melaphyres and basalts. To this can be added the calcium magnesium silicates – diopside and augite, both of which contain iron and appear in the same rocks as well as in crystalline schists and greenstones. Aegerite, a sodium iron silicate, which can form the finest hairlike, fibrous bundles, is found in granites, pegmatites and syenites.

The transition from gneiss to crystalline schists and greenstone schists is characterised by the amphiboles which include hornblende. They have a characteristically fibrous structure, often fine enough to be classified as asbestos. Anthophyllite, found near Heřmanov in the Vysočina region in Moravia, forms striking yellowish-grey radiating fibrous bundles in mica balls, and in the Transvaal it is mined in large quantites as fibrous asbestos. Anthophyllite is a magnesium iron silicate. In crystalline schists we find actinolite, tremolite, nephrite and grunerite. Actinolite is one of the most widespread amphiboles with its beautiful dark green, radiating bundles of acicular cystals. It is found in the talc and chlorite schists of the Zillertal, Austria. It also forms actinolite schist and, as grass-green smaragdite together with blood-red garnet, gives the unusual eclogite rock its unique colouring. Actinolite is also the mineral that forms the semi-precious stone amiant (hair-like actinolite) as well as the asbestos minerals mountain wool (byssolith), mountain leather, mountain cork and rock wood.

Hornblende (a calcium magnesium iron silicate) as well as occuring in crystalline schists, is also found in the transition rocks between hornblende granite and true greenstones. The sodium hornblende (arfvedsonite), with an iron content of up to 30%, is sodium iron silicate which, together with the sodium feldspar nepheline, forms many syenites and similar rocks. Closely related to arfvedsonite is riebeckite, whose tourmaline-like crystals point to this gemstone's close connection with hornblende. Not unlike riebeckite is crocidolite. It only occurs in crystalline schists. Its varieties include an asbestos and the two semi-precious stones tiger's eye and falcon's eye. The former is yellow, the latter a dull greenish blue.

The remaining mineral of this group is olivine, a magnesium nickel iron silicate, which independently forms rocks such as dunite and periodotite (of the greenstone group) as well as constituting a large part of gabbros, diabases, melaphyres, basalts and many crystalline schists. Olivine is one of the characteristic minerals of meteoric stones. In gem quality it is chrysolite.

It is typical of these mica, pyroxene and amphibole groups that magnesium and iron play major roles. Magnesium expresses the plant-forming forces; it is the main agent in the production of chlorophyll. Iron is present mainly in its ferrous form and gives these minerals their lighter or darker green shades.

The structure, colour and composition of these minerals taken together make it clear that plant and plant-animal forces were active in the forming of the greenstones and crystalline schists. The animal formative forces show themselves in the gradual introduction of sodium and calcium.

Iron ores constitute between 4% and 5% of the total terrestrial rock. Most of the ores are iron oxides, the remainder are pyrites and iron silicates. Iron oxide ores include magnetic iron ore (magnetite), titanium iron ore (ilmenite) and chromic iron ore (chromite).

By far the greatest quantities of iron oxide ores are to be found very finely distributed in the greenstones, especially in gabbro, diabase and the basalts. Magnetite is what gives the dark, almost black colour to many of these rocks, especially basalt. Iron ores are the only metal ores which take an active part in the forming of rock and thus indicate that iron plays a role as important in the Earth as a whole, as in the living plant, animal and human realms.

The increasing appearance of iron in the transition from granites to greenstones is closely linked with the early differentiation of life. Iron stimulates the individualising of forms in the kingdoms of nature.

14. Metals in the Life of the Earth

Whoever goes about with an observant eye cannot fail to notice that soils and rocks often have quite strong colours. Although the brown and yellowish shades of the weathered layer predominate, yet, if one penetrates deeper, all other colours are to be found. These colours are obviously not as vivid as the pure mineral colours such as cinnabar (vermilion), cobalt blue and chrome green, but on large sunlit surfaces they can lighten up more brightly. Besides the red sandstones and the striking red keuper marl, there is the rich colour range of the granites – blue from the Kösseine in the Fichtel Mountains, pink from Baveno in Piedmont, Italy, green from the central Alps and red from the mountains of Scandinavia. In the Grand Canyon of Arizona, the evening Sun touches with magic the multicoloured rocks ranging from gold to purple calling forth ever new transparent hues in the waning light. If one asks what causes these colours in rocks and stones and soils, the answer is always iron.

Iron with its colours penetrates the whole Earth. In the old rocks it is often so extremely finely distributed in its compounds with silica, oxygen and sulphur that it appears as if a mighty cosmic breath of iron impregnated the still airy-fluid living Earth. The Earth inhaled this iron and incorporated and transformed it in manifold ways, thus allowing different colours to appear. The earliest iron that the Earth took in is responsible for the greens and blues. In some granites, but above all in greenstones, we have this 'green' iron. Seen from this aspect, iron is strongly akin to nickel which also gives a green colour to minerals when present in very small quantities. Then there is chromium which also produces shades of green in the gemstones emerald, jade and nephrite. All three metals are found together in greenstones but it is the iron process that is responsible for their colour.

In older greenstones, some of which resemble granite, and in some schists as well as in basalts (descendents of greenstones), iron is found not only as green iron silicate but also in the form of magnetic iron ore. The latter contains much oxygen. It is a remarkable combination of ferrous and ferric oxide – a half burnt kind of iron. We should not imagine that this ore has arisen from red hot metallic iron. When the blacksmith draws glowing iron from the fire and forges it with his hammer, the little fragments that fly off (hammerscale) resemble chemically this half burnt iron, but magnetic iron ore (so named because in its natural state it is magnetic) has not arisen in a forge. Such metal ores in the old rocks point to living processes which formed them. The iron that the Earth inhaled was taken up by these life processes. Human beings, too, inhale minute quantities of iron with the air. It is most prevalent in late summer and autumn when meteor showers brighten the night sky. This is why the iron content in human blood begins to increase towards winter time and in the depth of winter contains far more iron than at midsummer.

The iron which we inhale and absorb into our blood is what makes blood flow. The blood-producing vessels are in the marrow of certain bones. Small cells detach themselves from the walls of these vessels. At birth the cells are colourless and possess a nucleus. As soon as iron enters these cells their nucleus disappears and they form red blood corpuscles. This primary blood is however a sort of virgin or primal blood. It has not been through the lungs or the liver. It is as yet neither blue venous nor red arterial blood. The bluish venous blood is a plantlike blood – it is saturated with carbon dioxide which it releases in the lungs in order to take up oxygen and become bright red arterial blood.

In venous blood iron is present in the 'plant-related' bivalent (ferrous) form, corresponding to green iron silicate. In arterial blood the iron is red owing to the oxygen and thereby so transformed that in this breathing process a soul-endowed being can dwell. In other words, the iron has become 'animalised'. In the mineral kingdom 'red' iron is always trivalent – ferric – to put it in chemical language.

Ferrous iron oxide in magnetic iron ore corresponds to venous blood; ferric iron oxide to arterial blood. Thus in magnetic iron ore we have a *mineral* form of iron corresponding to what, in humans and animals, is the iron of the primal blood of the bone marrow. Primal blood is neither venous nor arterial; it is the precursor of both. Magnetic iron ore occupies a similar position in the life of the Earth: it is the origin and beginning. It is the principal and oldest iron ore, and is very finely distributed throughout many rocks.

In the same way as the newly formed iron-enriched blood enters the arterial and venous life cycle, so we find in the subsequent development of rock, the appearance of red iron-stained sandstones and red iron ore (haematite). All this iron is trivalent iron oxide. It is combined with oxygen (like arterial blood). At the same time the bivalent form appears combined with carbon dioxide in the puzzling siderite or iron carbonate (like venous blood).

It may seem surprising that we try to connect the origin of such a mineral as iron ore with life processes of a higher order. But it is a continuation of the line we have taken from the beginning in observing the world of rocks. This particular comparison applies to iron because it occupies such a central position in all life processes. Iron is involved in the whole of organic development. The formation of its ores and all its other occurences in the mineral world have been shaped by life. Even the natural processes involving iron such as weathering and the infiltration of iron into springs and the living humus of plant-sustaining soils can be seen as life processes of the whole Earth.

The Earth has inhaled and incorporated this metal in order that otherwise invisible life should be made manifest in all its diversity of individual forms.

There are other iron ores which also play a part in rock formation. These are the sulphur compounds of iron – pyrite, marcasite and pyrrhotite (magnetic pyrites). Pyrite is mostly found in the crystalline schists and the blue roofing slates of Paleozoic time. It is also found finely distributed in granites and in fissures of the oldest rocks. This again goes to show how this heavy iron ore was suspended in the primal gelatinous state and then gradually crystallized in small crystals. It also shows that iron did not rise up from the depths but that it originated from the surrounding spheres. This is also why practically no workable deposits of iron ore are found in the oldest rocks. Iron originally entered the atmosphere in such a dispersed state that large quantities could not accumulate. However, in that atmosphere, which we described earlier as the world of the mineral-plant, there was sulphur – a substance which is not to be found in our present-day atmosphere. Had it remained in the Earth's atmosphere, subsequent higher forms of life could not have developed. Where did this sulphur come from?

All protein contains a small quantity of sulphur, but the primeval albumen contained sulphur and all the minerals found in rocks today. As the old life began to die, the albumen broke up; minerals were precipitated and sulphur dispersed into the atmosphere. It would never have disappeared from the atmosphere of the Earth if metals had not come down to the Earth from 'outside' and drawn the sulphur down with them. From this process all metallic sulphides originated. Pyrites, galena, cinnabar, etc. did not appear at once in crystalline form, but passed through the gelatinous state. Rudolf Steiner called this the 'parent state' of the later ores. One should not imagine that the process of sulphur combining with metals was merely a chemical action in the modern sense. On the contrary, it was an organic life process of the Earth.

The whole process presupposes a condition of metals that one might describe as aeriform. In the colouring of gemstones and other minerals we find evidence that in these early conditions of the Earth metals were indeed so finely vaporised that one could call them gaseous.

Rudolf Steiner described very early Earth conditions in which metals floated round the Earth like coloured clouds and gradually condensed from this airy state. We have the history of this recorded in the colours of precious stones. The colours are due to finely distributed metals especially iron, manganese and iron-related chromium and titanium.

Metals are rarely to be seen in their pure state except gold and copper. All other metals are not so intensely coloured as these two. However it is possible to reduce pure metals to fluid or vaporous conditions, when, as colloids, they show intense colours. In this way the metals have in a sense returned to an earlier condition.

Thus everything metallic is born of the air. If it is asked from where the metals in this aeriform state descended to the earth, we can refer to the experiments of Lili Kolisko. For many decades she studied the behaviour of metals by allowing solutions of metallic salts to rise up strips of filter paper during various planetary aspects (conjunctions, eclipses, oppositions, etc.). From the resulting chromatograms it was shown that certain metals relate to certain planets. It appeared for instance that when there was an occultation of Mars by the Moon, the solutions of iron sulphate and silver nitrate behaved differently and showed different pictures from those obtained when the two bodies were separated in the heavens. Thus it was possible to determine that lead was connected with Saturn, tin with Jupiter, iron with Mars, gold with the Sun, copper with Venus, mercury with Mercury and silver with the Moon. Some of these experiments have been reproduced by other researchers, These experiments show that when the metals are in solution they again come under the influence of the related planet.

Solid metal or metal ore does not reveal its origin. It is, so to speak, frozen, and rests in the ground like a seed. But the moment it is dissolved as a clear solution or a colloid, it becomes sensitive to the planetary sphere to which it belongs. If conditions in this sphere change – visible as astronomical planetary aspects – then variations also occur in certain chemical characteristics of the appropriate metal solutions on the Earth. This demonstrates scientifically for the first time the correspondences between metals and planetary spheres. The connection between metals and planets has been human knowledge through the ages. For ancient Indians, Persians, Egyptians, Babylonians and Greeks it was a fundamental fact. Rudolf Steiner confirmed it from his own research and encouraged experiments of the kind undertaken by Kolisko where these things could be made manifest.

How could it have happened that metals could reach the Earth in an aeriform state since the various heavenly bodies of our planetary system are so distant from the Earth? Let us remember how in earlier times and under past conditions, the Earth was far larger and contained within it all the other bodies of our solar system. At the beginning of the Earth evolution this combined body occupied a space extending to the present-day orbit of Saturn. At this stage it was a pure warmth organism, which had begun to differentiate. When this stage was completed, the Earth body gradually contracted to the present orbit of Jupiter leaving behind the planet Saturn as a separate body, and the Earth became a planet of heat, air and light. As this second stage drew to its close another shrinkage took place which left behind Jupiter on its orbit. At a third stage Mars was left behind. Later on in the course of the Earth's development the Sun was separated and the inner planets Mercury and Venus separated off from the body of the Sun. Meanwhile the Earth had shrunk to approximately the orbit of the Moon. Only after the Moon also withdrew from the Earth could that layer of rocks and minerals arise which is the subject of this book.

This is a brief sketch of the cosmology of our planetary system which Rudolf Steiner has elaborated in detail in the two books previously mentioned – *An Outline of Esoteric Science* and *Cosmic Memory*.

The extraordinary complexity and diversity of the creation of the world can only be touched on here. We have introduced it in order to show that the planetary system is an organic development and not a spiral nebula spun by an imaginary *deus ex machina* in order to throw off the planets as glowing clouds of gas as maintained in the Kant-Laplace theory. We present this picture of an organic cosmology in order to illustrate how metals in their primal form were carried to the Earth from outside in an aeriform state; how the withdrawing planets left behind, as it were, seeds in the Earth, and these seeds were and are the individual metals. Seen in this light the metals are gifts of the various planets and their existence in the Earth is like a memory of past stages of life.

Naturally many objections can be raised. One might say that if the metals have been built into the Earth in such a beautifully ordered sequence, then they should also be found in an orderly fashion in the Earth's strata. Another will say that there are far more metals than the seven here mentioned as connected with evolution.

To the first objection we would reply that this could not have happened since the Earth was and is a *living* organism and the metallic nature was incorporated as a *living* and not a mechanical process. The Earth had to digest the metals so that these gifts of the planets could be used by the life processes. Through the absorption of metals there arose an organic growth as in plants or animals, not a mechanical arrangement.

To the second point we would reply that the seven fundamental metals represent something like the seven notes of a scale. As there exists a great variety of intermediate tones within the scale so one can recognise intermediate tones between the metals. The seven fundamental metals are a harmony that was created by the pure planetary influences on the Earth. Subsequently the planets, between themselves, entered into relationships, aspects, (trine, square, opposition, conjunction, etc.). These accordingly produced new variations of joint planetary influences and through these other metals arose.

It is possible to determine the relationship of these other metals with the pure planetary metals by the way in which they were and are linked with certain organic processes.

When one considers metals as we have done; as something born out of light and air, one can comprehend Goethe's maxim that colour arose in the world as 'the deeds and sufferings of light'.

During the last decades ever new and surprising discoveries have been made concerning the connection of metals with living substances and organic processes. In the course of time it has emerged that among the heavy metals there is hardly one that is not found in small quantities in coal and oil besides playing a decisive role in the life of plants, animals and man.

We will select a few examples. Firstly there are the metals related to iron. For instance, manganese is very widely distributed and appears in small quantities in the oldest rocks. In schist and slate formations it forms ore deposits as does iron. Iron-magnesium silicates, which we described as rock-forming minerals in Chapter 13, are particularly rich in manganese. When we remember the life processes underlying these slates and greenstones belong really to plant life, then it seems that in the past manganese must have been a sort of iron for the plant world. Manganese has a peculiarity worth noting: in its natural state it is practically never compounded with sulphur (as iron is) but almost exclusively with oxygen. It also combines with silica as rhodonite and with carbon dioxide as rhodochrosite.

The relationship that a metal has to other substances in nature gives some indication of whether it was connected with life, and therefore can connect with it again, or whether it is something cast off from the processes of the mineral-plant. One can nearly always say of metal ores which are combined with oxygen that they are cast off from vegetative processes. On the other hand the sulphur ores indicate an all-embracing life process and have a refining effect, preparing the ground for higher forms of life. We mentioned this earlier in connection with iron pyrite, etc. The silicates and carbonates of metals also express a more plantlike nature just as the oxides do.

No wonder then that the organic remains of the Earth's past – asphalt, oil and coal – are rich in manganese. The massive deposits of manganese are to be found, as we said, in slates and younger strata, deposited while the old life increasingly died away. We find manganese as pyrolusite and psilometane always associated with limonite (iron oxide). These are oxidic ores.

Manganese is necessary for many plants today. Manganese deficiency causes diseases in certain plants and prevents the flowering of tomatoes.

All living organisms form specific enzymes and similar substances which play a mediating role in the processes of digestion and breathing. In these there is always a complicated protein base combined with a metal. They are true metal-protein compounds whose metal content develops quite astonishing functional properties in connection with the transformation of other substances. Here undoubtedly we meet that former life condition when everything mineral was embedded in protein. Later individualised forms of life have retained something of this as the basis of their most vital life processes.

It would be quite wrong to imagine that a metal-protein compound possesses some inorganic chemical properties whose interaction produces, so to speak, life functions. This is not the case. What happens is that the organic protein acts as a sort of window for the forces that stream to the Earth from the surrounding stars and planets. Through the metals being in suspension in a fluid state (a colloidal sol) in the protein, they become sensitive. What happens can be compared with the experiments of Lili Kolisko. The metal embedded in living protein again becomes sensitive to the influence of the planet from which it originated. In this way, through metals, the living protein is opened to the in-raying forces from the surrounding heavenly bodies. The starry heavens influence life. But also the life and soul forces of the individual organism can work on its own bodily functions through these embedded metals.

When we think of metals as originally part of and indispensable to life, we begin to find solutions to many problems that are difficult to explain from a purely scientific point of view. There is a small number of metals that are more plentiful than, for example, antimony, uranium, silver, bismuth, mercury, gold and platinum, which are not found in concentrated deposits of ore but are so finely distributed that it is not possible to extract them in any quantity. They were discovered by spectroscopic analysis. One of these is scandium, a rare earth of which one finds infinitesimal traces in nearly all minerals which constitute stanniferous granite. It is also in all clay minerals and is easily detected by the spectroscope. Scandium was therefore already present in very early times of the Earth when granite began to form out of the mineral-plant. As we have said before, this mineral-plant life was of a low order and relatively undifferentiated. In a certain way it could be compared with our smallest present-day soil fungi – a form of life that has not developed further but remained stationary. Scandium was obviously important for these low forms of life, and indeed it has been found when studying the biological properties of scandium that simple moulds like aspergillus cannot live and grow without it. When we consider how such moulds seem to grow everywhere, we can appreciate the widespread distribution of scandium.

The metal gallium is another example. It is found in zincblende (sphalerite) and grey copper ore (tetrahedrite) in sufficient quantities to be extractable. All clay minerals contain this metal. Its properties, like those of scandium, point to a relationship with tin, zinc and aluminium. Since aluminium silicates were the most important substance in the life of the old mineral-plant, it is understandable that gallium is vital for the existence of all plants today. If gallium is carefully excluded from growing plants the yield is reduced by up to 40%.

Another metal that has come to the fore with the development of transistors and electronics is germanium. It is found in tin ores and copper deposits, but the largest quantities of this valuable metal are obtained from coal. It must have played an important role in the fleeting plant forms and life processes of Carboniferous times. It is sufficiently abundant in coal to be extractable from the ashes.

Even in the old coals that precede the true Carboniferous period one often finds unusual concentrations of metals which show that these must have played an important part in the life processes that produced the coals. Thus we have, associated with Canadian pegmatites, a dense bituminous coal known as thucholite which contains considerable amounts of radioactive uranium and thorium. Similarly there is a very old Swedish coal which, besides radioactive uranium, also contains metals such as nickel, copper, zinc, molibdenum, vanadium, lead, tin and bismuth in the ash. These old coals are sometimes found in granitic rocks. They belong therefore to the time when life was not yet differentiated.

Even today lower plants such as algae can store and 'digest' uranium and radium, and this makes it understandable that in early times quite other quantities of metallic substances were enveloped by life. Earlier it was assumed that this metallic and mineral content of coal ash was carried by water into the coal at a later date. It appeared unusual that such quantities should be present, for it was not known that metals were necessary for life. Meanwhile it has been found that these metals are also present in oil bound up with complex organic materials. Metals such as nickel, vanadium, copper and molibdenum are found in oil compounded with porphyrines. The composition of the porphyrines resembles that of red haemoglobin of human and animal blood and is also closely related to green chlorophyll in plants. This shows that in earlier times metals were connected with life to a far greater extent than now. The durability and persistence of metal-containing oil porphyrines is such that they have endured unchanged to this day and are not even affected by the high temperatures to which they are subjected in the processes of refining. They can still be found in petrols and lubricating oils where they can have detrimental effects.

Vanadium and molybdenum are necessary for the nitrogen bacteria on the roots of leguminous plants such as beans, peas, clover, and tungsten for other bacteria. This shows how vital the role of metals is even today, though no longer in the quantities of former times.

Another metal, quantitatively after iron, is titanium. The titanium content in rocks is about one tenth of that of iron. It usually occurs combined with iron as ilmenite ore but can also be found independent of iron associated with quartz as the beautiful rutilated quartz (Venus' hair stone). This shows that this metal occupies a position between iron and silica. Ilmenite (titanium iron ore), rutile and other titanium minerals usually occur in the oldest rocks which contain much silica. They are usually finely distributed and do not form ore deposits. Thus titanium can only be obtained when after weathering and transportation, its heavy particles sink to the bottom as ilmenite sand. These dark sands are fairly widespread on many coasts. Through its presence in the oldest silicic rocks titanium shows its connection with the life of the mineral-plant, and even today nearly all plant leaves contain titanium.

Chrome is also found with iron as chromic iron ore (chromite). It forms few ore deposits and mainly occurs finely dispersed in serpentine and other greenstones. It gives both rubies and emeralds their colours. Chrome is a hard metal and we find it in the densest and toughest rocks belonging to what we have called the tree stage of the Earth.

All these metals (which are related to iron) including vanadium, molybdenum, tungsten, cobalt and nickel are found in the oldest strata. They are present in perceptible quantities, combined with proteins, in the body fluids and organs of living organisms where they are known as trace elements. One can perceive the effect of minute quantities, but these effects cannot be explained purely chemically. The significance of these effects is due to their fineness and dilution, that is, to their enormously increased surface area. Forces then come into operation which have nothing to do with ordinary chemistry.

If we go a step further we reach a domain not known to modern science and we come to the effects of those metals we have not yet dealt with, namely lead, tin, gold, silver and mercury. These metals, like iron and copper, are found in fairly plentiful deposits. Only gold tends to be finely dispersed. To understand the role of these metals in the life of the Earth we need to know something of their planetary origin and their relation to the inner organs of higher forms of life. Fundamentally the planetary metals should be considered as *organs of the Earth* and the large ore deposits in the Earth recognised as the dead remains of these organs. Once upon a time they were a living, in-streaming activity of spirit-filled creative powers from the surrounding cosmos which then coagulated to dense earthly matter. Wherever we find large quantities of these metal ores concentrated and embedded in the rock, we must picture in former times mighty planetary forces streaming in from outer spheres. Thus, in coloured clouds, the metals came to the Earth. Later they condensed to the liquid state and flowed into the fissures of the Earth. This explains why all large metal deposits peter out as one follows them deeper into the earth. It is wrong to think that metals came up from the interior of the Earth. They came from outside as meteoric iron does to this day.

It is a mistake to draw conclusions about the constitution of the Earth 6000 km (4000 miles) to its centre when one only knows a few kilometres of the crust.

The sparse distribution of metals in deep-seated rocks and the sudden petering out of nearly all sizeable ore deposits at depth suggests quite another picture. Originally everything metallic was in an airy state around the Earth. As it streamed in, in the early times of the Earth's life, it became life processes and was finely distributed in the rocks. Only later as life began to differentiate – in the Schist and Porphyry period – did the metals form ore deposits. This is why the major deposits coincide with the transition from primal rocks to slates, schists, greenstones and porphyries. The 'organs' of the Earth were first formed in the fluid-living state. Then lead acted on the inner framework of the Earth, tin created the correct balance between solidity and fluidity, gold with its urge to disperse held sway between gravity and forces of lift in the living strata of the old atmosphere, silver regulated heat, and mercury the circulation of fluids in the living Earth.

All living organisms have retained from the past something from the living influence of metals which extends beyond mere crude substantiality. To understand the life of metals in the creation of the Earth, we must look at their action in humans, animals and plants. Here they are still as active as they once were in the Earth as a whole. Recognising the effect of trace elements is a step in the right direction. These traces, however, conceal even greater secrets which lie behind the crude mineral matter in the creation of man, animal and plant. Therefore a future science will attempt to solve the riddles of the mineral world by studying the embryological development of living organisms, especially of human beings.

15. Embryonic Development and Rock Formation

The innermost purpose of creation from its earliest beginnings was the development and emergence of the human being. What we see around us as the mineral, plant and animal kingdoms are *not* the ancestors of present-day man, but are the cast-off husks of an all-encompassing being that had to lay aside its sheaths in order to free itself from the compulsions of its own evolution. Nature arose in order to give a spiritual being the possibility to rise from its dependence on matter and seek in freedom its spiritual origin. Yet out of the creation of the world as a totality there emerge human beings who, in so far as they are beings of nature, are able to forget their spiritual origin. Since they are aware of their natural self, they are placed in freedom to remain creatures of nature or to progress towards developing their spiritual self. If human beings do the former, they make what should be a stage of development, into a fixed condition and separate themselves completely from the evolving world. This leads to hardening, death and self-destruction, as the present state of humanity suggests only too clearly. If however they recognise themselves as spiritual beings and overcome and transform their natural state, they return to their source as *free* spiritual beings to take their place among the creative and progressing powers of the universe.

The human physical body submits to the inexorable laws of birth and death, growth and nutrition. This body is permeated with solid mineral substances, fluids, air and also heat. As we have seen, these four elements are also the stages of the coming into being of the Earth. The beginning is not the creation of visible, mineral matter, but is *heat*. In the primal beginning of creation the human spirit was embedded in warmth, its body was flowing streams of warmth. This body had neither life nor awareness, nor did it offer the possibility of developing a consciousness of self. The present-day mineral is at this stage of consciousness. The whole world of rocks has remained at this stage, left behind, one might say. Rudolf Steiner called this first evolutionary stage of Earth Old Saturn.

The next stage, called Old Sun, brought life to this body of warmth, and with it the formation of air and light. Human beings became plantlike with a body of warmth and air. The whole plant world of today is at this stage. Thus a part of what was originally created moved on a step, and part remained behind. This last development was preceded by an abbreviated recapitulation of the earlier stage in a transformed condition.

At the third stage, called Old Moon, the human body was endowed with awareness and now consisted of warmth, air and water. Besides the fluid element, sound came into being in this Moon-Earth. Human beings advanced to an animal stage. What remained behind at this stage later formed the animal kingdom.

As human beings entered into the stage of the real Earth, they first had to recapitulate the previous stages in order to incorporate solid mineral matter into the physical visible form of today which became the foundation of awareness of self.

In these recapitulations we have the familiar biogenetic law propounded by Ernst Haeckel which states that the organic development of an individual organism is a contracted recapitulation of the history of its species. The history of the human species is not the development from bacteria to man, but from a spiritual seed surrounded by warmth to the present mineral-filled visible form of the human body, able to be the bearer of a spiritual being conscious of itself. The human body is the oldest and most perfected possession of humankind. When the human individuality prepares to re-enter earthly existence, it needs the help of all the creative powers which have from the beginning taken part in the creation of the human body. In the building of this body, enveloped in the mother's womb, the whole evolution of humankind and Earth is recapitulated. We can follow the processes of the evolution of the Earth in human embryonic development, recognising three stages.

The first is the morula stage where, from the fertilised unicellular egg, a globular structure somewhat resembling a mulberry (morula) is formed. It has neither inside nor outside: one might describe it as granular – formed out of many equal cells.

The second stage is that of the flat surface of the germ layer stage when the differentiation of organs begins. By invagination and upward surging the cavities of the digestive tract and the spinal canal are formed.

The third stage begins with the depositing of calcium in the cartilaginous skeletal system, in other words, ossification.

These three stages of early embryonic development up to the onset of ossification of the skeleton, as regards structure and form, reflect a faithful picture of the basic structure of the world of rocks as we described it. There is the granular structure of the primal rocks, especially the granites, which is followed by the flat layering of the schists and slates, which are immediately acted upon by the cavity-forming processes of porphyry; then calcium gradually infiltrates, becoming predominant in the Jurassic period.

It is important to realise that in this parallel, human evolution is the primary phenomenon, and events in the world of rocks are the results. This shows the rocks from a new point of view. They have an intimate relation to the development of the human form. They are the cast-off husks and residues from its formation.

Human beings, however, are not only a form, they also have life, awareness and conscious thought. One can ask: where are the cast-off husks from the development of life and awareness to be found? We have to remember that during development human beings also left behind the plant and animal kingdoms. As the rock kingdom is related to the development of human *form*, so the plant kingdom is to the development of human *life* and the animal kingdom to the development of soul and *awareness*. In the realm of individual conscious thought, however, man rises above the kingdoms of nature to what is specifically human.

To return to embryology and rocks – we spoke of three processes: morula stage, germ layer stage (differentiation of organs) and the beginning of calcification. Threefoldness is also to be found within the second stage of the embryo. Three different layers develop – the outer germ layer or ectoderm, the middle layer (mesoderm and mesenchyme) and the inner layer or endoderm.

Each of these layers takes part in the forming of certain parts and organs of the human body. The ectoderm is the origin of the greater part of the sense organs, the skin, nerve cells, and also the lenses and vitreous humour of the eyes. All these have a special relationship with silica processes in rocks, particularly evident in granite. Considerable parts of these organs contain silica or the silica processes play an important role in their functions.

The middle layer is mainly concerned with the creation of the organs of breathing and blood circulation – heart, blood vessels and blood cells. Apart from the lenses and the vitreous humour of the eyes, it takes part in the building of the predominantly skinlike surfaces of all organs. The membranous creations of the mesoderm are the human schist/slate process. The mesenchyme part of this middle layer is the builder of the supporting and filling-in tissues between what is formed by the other layers; hence it is also responsible for the basic tissue of the bones.

The calcification of the bones originates from the inner layer (endoderm) which forms the nutrition system of the developing organism. The membranous lining of the whole digestive tract proceeds from the endoderm.

In the mesenchyme part of the middle layer lies the origin

of the important process of cavity formation; it also helps to give structural and functional details to the various organs. All organs embodying cavities, including the bones, proceed from the mesenchyme. We have the counterpart of this embryonic process in the porphyry formations of the rocks.

Through the activity of the entire middle layer – forming surfaces and cavities – a system of organs develops that is a reflection of the interplay of the surrounding planetary system. In their development human beings have taken into themselves all that is around them. Not only do they partake of the lower kingdoms of nature but also of the surrounding world of stars. We can only touch on this here. Rudolf Steiner has dealt with it extensively in his writings and lectures.

The inner human organs correspond in a certain way to what the metals are in the Earth, which, as we have shown, originate from the planetary spheres. Just as in humans these organs originate from the middle layer (corresponding to the schist/slate and porphyry processes in rocks), so we find the main deposits of the seven planetary metals in the schistose and porphyritic rocks of the Earth. We will not find substantial metal deposits in the typical granites or in the Jurassic limestones. It is where granite begins to give way to gneiss and schist or in the early porphyries (pegmatites) that the metals first appear in seams, deposits and veins. In the real primal granites the metals are extremely finely dispersed, as we have seen.

In true limestone formations, the metals again recede and we only find them when limestone is interlayered with slates, sandstones or volcanic rocks. In most cases however, these younger ore deposits are of a secondary nature; that is, they have arisen through the disintegration, dissolution and re-deposition of older metal-bearing strata.

With these younger limestone deposits we touch on a process outside which, inside human beings, is connected with the origin of the digestive organs during the germ layer stage of the embryo. We have seen that the membrane lining these organs originates from the inner layer or endoderm. This applies especially to the alimentary canal and its glands, the inner tissues of the liver and pancreas, the inner lining of the lungs and part of the organs of taste. In the forming and functioning of these organs, the nourishing and building-up forces of calcium are active, but they do not here reach the point of forming and depositing calcium substance.

The nature of calcium on the one hand has nutritive and up-building forces, and on the other hand, bestows the substance that makes the human skeleton. The connection of calcium with the animal world is evident considering that the major deposits of calcium in the Jurassic and Chalk formations are due to vast quantities of shells of small organisms arising from the tremendous flooding life of the Earth's past. Such abundance of *formed* life is only possible through the excessive nutritive and up-building forces of calcium. The older life of the Earth with its mineral-plant and plant-animal, predominantly dependent on silica, was not like this and left no concrete forms behind – only impressions of attempted forms. From silica processes of the past the formative forces working on the periphery of the human body have developed, while the inner structure is due to calcium processes. It is of particular interest to note how calcium is transformed when it enters the skeletal structure. The calcium which forms our bones is not calcium carbonate like the limestone mountains, but mainly calcium phosphate. Calcium carbonate has more of a connection with vegetative life processes while calcium phosphate provides evidence of the active nutritional process of the animal organism that took hold of the calcium. Phosphorus plays a most important part in our whole metabolism. This is noted in passing and cannot be elaborated upon further at this point.

The ossification which begins from the nutritional function of the endoderm does not proceed uninterruptedly. Regarding the very complex process of ossification of the skeleton certain facts have recently been established which support the principle of recapitulation. It has been found that in the embryo this process is initiated four times, and three times it retreats completely. Ossification begins three times from different classes of tissue (cartilage or connective tissues), and before each new beginning the calcium is completely withdrawn and only at the fourth onset does it proceed uninterruptedly to post-natal time.

This fact – incomprehensible from a utilitarian point of view – can be explained if one sees, in this fourfold repetition, a picture of the whole creation of human beings and the Earth. What took place in the stages of human creation, from the embedding of the human spirit-seed in warmth, to the body woven of air and life, to the fluid body imbued with awareness, up to the present body permeated with mineral substance which is the basis of present-day consciousness and the power of thought – all this is reflected in embryonic development.

This approach to the study of geology based on human development – only briefly outlined here – is not only the most important but also the most difficult. Its importance will be appreciated if, in the future, the animal and plant kingdoms are thought of as linked with human development. In the combining of scientific data with the results of spiritual investigation lies the universality towards which so many strive. Creation itself is a unity, it is directed towards one goal – humankind.

16. Time in the Development of the Earth

In the previous chapters nothing has been said about a factor that is of great importance in modern geology, namely – time.

Human thought has derived its measure of time from rotation of the Earth, from the circling of the Moon round the Earth and of the Earth round the Sun. The fact that the regular repetitions of these motions, through thousands of years back to the earliest known civilisations, tally with our present-day calculations, has led, with some justification, to the idea that these familiar rhythms of time – days, months and years – hold good for millions of years before and after the present time.

However, besides this perfectly logical conclusion, there is another, equally valid. It is that all repetitive occurrences point to something living, developing, ever-renewing itself. In contrast, the mineral kingdom – crystals, ores and rocks – which is cast off from life, presupposes duration – a tendency to persist in its form. This persistence is to some extent an illusion, for, as we have seen, in weathering and dissolution mineral material can be led back into the stream of life. On closer investigation, as we shall endeavour to show towards the end of Chapter 17, this 'duration' of the mineral is part of a far greater rhythm than is usually imagined. There is nothing really dead in the world, only varying cycles of becoming, existing and passing away.

Nonetheless we can say as regards the present age: Life exists in *time* and what falls out of life exists in *duration*. Strangely enough though, even in the rhythms of our time there is an element of duration in so far as the recurring of these rhythms is calculable. Yet ultimately, all rhythms of living entities are incalculable.

From the fact that what we call time – determined by the relation of the Earth to the surrounding stars – has a certain periodicity, it follows that all the participants in the rhythmic phenomena stand in a living relationship to each other, so they must be themselves living entities.

All living beings are the result of a development from the past into the future, and have, in the early stages of their existence, different rhythms of growth from those in middle and later life. One has but to compare the pulse and breathing of a baby with that of an adult and of an old person.

As we have seen the Earth has passed through stages of a development arising from creative deeds of spiritual beings. These stages took their course in a cosmos which did not have the settled state of our present world. To enable the present order of our solar system to arise, one by one the planets and finally the Sun and Moon separated from the developing Earth (which included humans and the emerging kingdoms of nature). Thus, up to the separation of the Moon, time could not be reckoned in our present years since the Moon as a heavenly body was not yet incorporated in the system. Only after this event could our planetary system begin to assume its present order and rhythms including its relation to the zodiac. Between the separation of the Moon and the present, time periods approximated more and more to our present years.

In this connection Rudolf Steiner repeatedly observed that vital turning points in evolution, such as when time in the modern sense of calculable rhythms began, are always 'connected with certain positions of the heavenly bodies'.* Naturally such aspects could only occur after the Sun had separated from the Earth and begun its course through the zodiac. However, Sun and

---

* *The World of the Senses,* lecture of Dec 31, 1911.

Earth did not immediately after the separation of the Sun follow a predictable and precise path. One can say it took thousands and thousands of years (though not measured in years, since they did not exist) till day and night periods and yearly rhythms harmonised. To complete the organic process towards calculable time the Moon had to separate from the Earth and start on its own orbit, exerting its influence from outside.

For those interested in the astronomical aspect we recommend Elisabeth Vreede's *Astronomy and Spiritual Science.* This work, which elaborates Rudolf Steiner's findings, deals with the precession of the equinoxes, aphelion and perihelion, variations in the eccentricity of the Earth's orbit, obliquity of the ecliptic (angle of the Earth's axis), among other things. From this it appears that reckoning time in years in a universe moving according to mechanical laws is possible from around 20000 BC. This point in time was approximately the middle of the Atlantean epoch and the middle of the Earth evolution. Some 4000 years later, around 16000 BC, the Ice Age began. This period of the Atlantean development was one of extraordinary climatic variations alternating between widespread glaciation and almost subtropical conditions. What occurred is not only due to celestial phenomena but also connected with the development of humanity. It is only a few thousand years since human beings in their present form walked the Earth. The prehominids which have been found in all continents were creatures whose forms hardened too early, but they have nothing to do with that part of humanity whose bodies did not become hard and solid until they could become apt vessels for spirit gifted with self-awareness.

The climatic alternations were the instruments of creation to either train humanity or wipe out those who were too weak.

During the Ice Age not only were the climatic zones greatly changed but the atmosphere also was completely transformed. In his book *Cosmic Memory,* Rudolf Steiner described how the Atlantean atmosphere contained far more water than today – the air was 'thicker' and the water 'thinner' resulting in a completely different interplay of the two elements. This produced the misty land of North Atlantis, whose climate was decisive for the development of a part of humanity. In Nordic mythology North Atlantis appears as Niflheim (home of mist) and the warmer Southern Atlantis as Muspelheim. The misty land of Atlantis did not know the rainbow because the Sun could not penetrate the thick mist. Not until the Flood of around 10000 BC and the end of the Ice Age, did water in masses pour down from the atmosphere, the sky clear and the first rainbow appear as recorded in the Bible.

From this it follows that there are certain limits to the validity of our calculations of time. Observing natural phenomena such as the depositing of mud and sand in lakes, river estuaries or bays – or even radioactive decay – it is important to be aware of the limits of the duration of our time scale if we are not to fall into the error of projecting the laws of the now ageing Earth into the time when the young Earth was coming into being.

It would be as if a scientist, after noting the pulse and respiratory rhythm of a grown man throughout one year, would proceed to make a graph, and projecting the curve deduce what his pulse and respiration would have been 300 years previously and would be 300 years hence. The observations of the scientist are unquestionably accurate and his calculations equally so, but pulse and breathing of the man do not continue for 600 years.

It is the same thing as when astronomers, basing their calculations on present-day slight changes in the precession of the equinoxes, the angle of the Earth's axis, aphelion and perihelion, which undoubtedly hold good for 25000 to 30000 years, project these rhythms without limits into the past and future, and therefore reach unsound conclusions.

In this way calculations have been made which go back 900000 years and produce a curve of the intensity of solar radiation. In the geological field it was possible to trace a corresponding series of glacial deposits for part of this time, which showed a remarkable parallelism. The parallelism is not questioned; only whether it can be measured in years.

Another example is reckoning the age of the Earth from the present-day rate of decay of radioactive elements. These calculations are based on the supposition that the decay of radioactive elements began simultaneously with the formation of these elements in the cosmic process of solidification. Therefore, since the quantities of the end products of decay in a given time are calculable, one can determine how old a particular mineral is, and thus how old the whole Earth must be. This results in ages varying from 200 million years for a uranium pitchblende from Jáchymov in Bohemia to 2600 million years for a monazite from a Zimbabwean pegmatite. Similar tremendous differences in the rate of decay of radioactive elements are found when calculations are based on contemporary phenomena. In this way the radioactive half-life (the time taken for the number of radioactive atoms of a particular substance to be reduced by half) ranges from millions of years in the case of uranium and thorium to fractions of a second for bismuth-214, a decomposition product of radium.

An extremely interesting phenomenon in connection with rates of decomposition is that the shorter the life of the radioactive element, the stronger the penetration of the alpha particles. Intensity of radiation stands in an unmistakable relation to time. The questions arise: why is the decay of different radioactive elements so variable, why do atoms only decay slowly as they give off energy, and why does this decay not occur all at once? There are no simple answers to these questions as long as one considers *naturally* occurring radioactivity only.

Humankind has now learned manipulate the time factor of radioactive decay, to speed up or slow down the natural process, or even to produce artificial radiation electromagnetically. Basically the processes have been brought under control by embedding large quantities of pure radioactive substances, such as uranium, in graphite or heavy water.

By contrast, the changes, which take place very slowly in nature owing to the fine distribution of these substances embedded in the rocks, can be tremendously accelerated and enhanced by bringing together large quantities of radioactive material leading to a catastrophic chain reaction.

So on the one hand we have the principle of the nuclear reactor producing energy from controlled decay, and on the other hand the nuclear bomb.

The intensity of decay depends on the concentration of the mass. The extremely slow, natural process of transformation of uranium-238 into plutonium-239 can be either directed to the production of energy over time within a nuclear reactor, or it can be made to take place almost instantaneously in a fraction of a second in the nuclear bomb.

A small quantity such as 200 grams of uranium-235 or plutonium-239 is harmless, but if it is increased to several kilograms (about the size of a coconut) the chain reaction sets in which leads to detonation. In practice this reaction is triggered off by bringing together two hemispheres of uranium-235 or plutonium-239. Thus it is clear that there exists a relation between the weight of pure radioactive substance and the time it takes to decay.

Since pure substances do not occur in nature, the intensification which can be obtained only from pure uranium or plutonium will never take place in nature. For rapid disintegration a relatively pure radioactive substance is necessary, so that little of the radiation connected with the formation of neutrons may be diverted by being absorbed into foreign stable material. Since in nature these substances do not occur in a pure state, such violent reactions will never take place.

It is conceivable that natural radioactivity has not always existed at the same intensity as today. The rate of decay of certain elements may have increased during the passage of time and not have maintained a steady rate from time immemorial. This may appear to be incompatible with the presence of lead and helium in radioactive minerals since these are presumed to be the stable end-products of decay as observable today. This seems to be borne out by the different atomic weight of this lead compared with lead ore found far removed from any radioactivity. It could however equally well be that the ubiquitous helium and lead were always present in these minerals. Thus it is possible that radioactivity is not of the great age generally surmised. Errors like this arise from projecting present-day phenomena into the past and the future.

17. The Earth as Seed of a New World

The solid element of our planet Earth – rock – appears to us as a realm apparently completely at rest and no longer in the process of becoming. The rocky summits of mountains have remained almost unchanged for thousands of years, although in further millennia weathering will reduce the created world to dust. The weathering is due to the elements from which life unfolds – water, air and heat. Life as revealed in the higher kingdoms of nature has nothing of the quiet and settled state of the rocks. It pulsates in plants between seed and fruit, in animals between birth and death and leads human beings beyond birth and death to find the way to their spiritual origin at a higher stage of existence.

The time in which the rocks arose came to an end thousands of years ago. Today the stage of life of the Earth is comparable to a ripe seed which has fallen to the ground and rests in the winter soil awaiting germination. This present stage is best expressed in such imagery taken from the plant world, for the plant seed is a completed stage which can lie dormant for shorter or longer periods. It is also something hardened and mineralised and contains only a minimum of life. Ripening is a mineralising microcosmic process which corresponds to the macrocosmic process of crystallising and hardening of the world of rocks. The hardening of the seed, like the hardening of the rocks, is the unavoidable pre-requisite for the emergence of a new plant or of a New Earth.

What takes place when a seed germinates? Water, air and warmth act upon the hardened seed breaking down its substances into non-organic substances. Only after this dissolution and reducing to chaos, can the constructive and revivifying formative forces of the plant take hold and call forth the seedling.

In one of his last Letters to Members Rudolf Steiner posed the question, what is the Earth, in reality, in the macrocosm?* His answer is that spiritual research reveals that the original state of the macrocosm was extraordinarily full of life which gradually died. The macrocosm became more and more subject to calculable laws. The life that died out of the macrocosm became microcosmic and is to be found everywhere where seeds germinate and embryos develop in nature.

The present Earth, with its kingdoms of nature and humanity, developed from the process of dying. The dying of the macrocosm was the necessary condition for the development of man as a being conscious of selfhood. Through these processes man and all nature carry something germinal into the future.

At the end of the above letter to members Rudolf Steiner called the Earth 'the embryonic beginning of a newly-arising cosmos.' In the microcosmic life of the kingdoms of nature there is today a superabundance of seeds and eggs which perish and one may well ask what becomes of these lost forces. According to Rudolf Steiner the superfluous seed forces of the plant world provide the *substance* for a new image of an ordered macrocosm. The superfluous forces of the mineral world carry the forces from the plants to their allotted places in the macrocosm. The superabundant germinal forces in the animal world act in such a way that all that is carried out from the plant world by the mineral forces into the universe is held together in a sphere so as to present the aspect of a macrocosm rounded in on all sides and self-contained.

Here Rudolf Steiner pointed unequivocably to germinal forces in the mineral kingdom. Where are we to find these if we look

---

* *Anthroposophical Leading Thoughts,* letter of January 1925.

deeper into nature in the light of spiritual cosmology? We have seen that all the rocks of the Earth initially arose from tremendous plant and animal life-processes. These were predominantly life-processes of the mineral-plant and the plant-animal. This can be recognised by the fact that over 90% of the rock mass is composed of silica minerals and only about 0.2% consists of calcium, of animal residues. Thus rocks are predominantly formed from vegetative life-processes. The young limestone strata are but a thin skin or shell patchily covering the Earth.

Everything resulting from dying vegetation contributes to a new life. This applies not only to seeds but also to fallen leaves and whole plants as they wither and die. The seed can become a new plant or it can, together with the rest of the plant, rot and be absorbed into the living humus of the soil. Under natural conditions the dying plant always contributes something towards the maintenance of the life in the soil. The germinal nature of this decaying process is not obvious and is only revealed in the formation of humus when the plant residues are united with the minerals of the earth. At first the plant tends to dry out and harden and then it disintegrates and provides the material for the activities of lower organisms. The resulting substances resemble inorganic matter without actually becoming mineral. These processes resemble fundamentally what happens when a seed is formed and, after the winter's rest, proceeds to germinate. As regards the substances present, there are of course great differences, but the *processes* are closely related.

When we transpose these ripening and dying processes of the plant to the macrocosmic scale of the whole Earth we approach the idea of the birth of rocks. Silica then played a similar role to carbon in the plant today.

The process of rock formation came to an end about the middle of Atlantis – some 15 000 years ago according to Rudolf Steiner.* Rocks at that time were not so hardened as they are

---

* *From Comets to Cocaine,* lecture of Feb 17, 1923.

today. The hardening process continued for many thousands of years. Up to the last millennium BC enormous buildings were erected by many different peoples using crystalline rocks which today are very hard. None of these peoples were acquainted with iron or steel, yet the blocks of stone were dressed with such precision that one cannot insert a knife blade into the mortarless joints. This suggests that the rocks were not so hard in those days.

Even today it is known that freshly quarried granite, sandstone, slate and limestone are softer and can be more accurately dressed than the same rocks after exposure to air for some time. Air and warmth dry out the rock and often cause chemical changes. One of the most striking examples of this hardening process today is emerald. In some localities these gemstones are found embedded in soft mica schists. On extracting them – in spite of their great beauty and transparency – they are often so soft that they can be crushed to a damp powder between two fingers. These soft crystals must be handled very carefully and have to be kept for some weeks in a wooden box to dry out slowly. They then become as hard as rock crystal. This is why most emeralds show a network of fine haircracks.

The hardening of rocks is not a continuous process but is rhythmic and cumulative. In one of his lectures to the workmen Rudolf Steiner said that the hardening began after the separation of the Moon in Lemurian times.* In certain areas the Earth solidified to the density of a horse's hoof and after a while became softer again. The hardening then began again in another area and so on. In a later lecture he described this rhythmic process from a different aspect, showing how the condition of the Earth is dependent on the course of the Sun through the zodiac.† Thus 25 920 years ago (one Platonic Year) the Earth was in a somewhat similar condition to that of the rocks today. Then as now the Sun

---

* *From Crystals to Crocodiles*, lecture of Sep 20, 1922.
† *From Comets to Cocaine*, lecture of Feb 17, 1923.

at the vernal equinox stood in the constellation of Pisces. This coincided with the end of Lemuria. The alternating processes of hardening and softening which set in after the separation of the Moon had, by the end of the Lemurian time, led to a condition which can be compared with that of today.

In the same later lecture we are told that in between, when the Sun (at the vernal equinox) stood in Libra, the Earth became again soft and pliable and was like a living plant. That was the end of the time when new rock arose, about 15 000 years ago, which we mentioned earlier. From this time on the hardening began which has now passed its peak in our time. After a Platonic Year of some 26 000 years, when the vernal equinox will return into Libra in about 11 000 years, the Earth will again be in a living plantlike condition.

From all this it becomes clear that the present mineral state of the rocks only exists for one Platonic Year. At the beginning of this Platonic Year when the vernal equinox was in Libra there was a balance between the fluid and solid states of the rocks. The hardening process has not completely penetrated the interior of the Earth even now, as we have seen by the effect of air on rocks when brought to the surface.

Before the rocks became so hard and dry that the elements of life – warmth, air and water – could no longer affect them, disintegration set in through weathering and other destructive forces. Natural radioactivity is of course one of these – the self-disintegration of the densest materials of the Earth. Radioactivity even has the power to reverse crystallisation and transform crystalline materials into gel-like substance (see Chapter 11). We are living on an Earth that is already on the way to the time when the vernal equinox will again be in Libra and the Earth will be in a state of balance as it was 15 000 years ago.

Mineral existence is therefore only a transitory stage. It is not a state of complete death but is a prolonged stage of subdued life which once again will awaken to full force, but transformed. We must now return to our comparison with the dormant plant seed.

In the Letter to the Members previously quoted, Rudolf Steiner emphasised the significance of the surplus germinating power in the kingdoms of nature. What flows out from the plant world provides, in a sense, the material for a new-arising picture of the macrocosm, of the future Earth. What comes from the plant will be ordered and moulded by the surplus forces of the minerals. And what flows out as surplus reproductive forces from the animal kingdom acts on what has gone out from the mineral and plant kingdoms and forms it to a sphere, rounding it and holding it together so that it presents the picture of a self-contained macrocosm. Rudolf Steiner pointed here to a macrocosmic law which throws light on many natural phenomena. It reveals the inner meaning of the interaction of the three kingdoms of nature. What is described as the working of the forces of these kingdoms of nature for the future of the macrocosm must also find its reflection in the microcosm. One may recall the old hermetic saying, 'As above so below'.

If one accepts this law one may well ask where it is manifested in the world of the smallest entities. Here the scientist's mind turns towards the manifold world of symbioses, commensalism and other intimate relationships which exist between the three kingdoms of nature – a world full of enigmas, whose laws can be stated but not explained.

There is a fundamental process in which the harmonious interaction of the three kingdoms of nature follows the macrocosmic law and can be seen reflected microcosmically. This is humus formation – the living soil in which plants can grow and which supports the higher forms of life. We described in Chapter 4 that this humus formation is a transformed process from the past. Here it can be shown how, from another aspect, this process corresponds to the macrocosmic law.

We pointed out that everything proceeding from the dying plant bears something of a possibility of future life within it. It can be considered as surplus germinating power. All that withers and dies has the potential of taking part in germinating life.

If plant remains are burnt, these potential forces unite with the picture of the new macrocosm. If, however, they are composted with soil as described by Rudolf Steiner in his Agricultural Course, the surplus forces of the plants are united with the forces proceeding from the disintegrating minerals. These forces resemble what takes place in germination. What finally results from the complete breakdown of mineral substance can be seen in the cloudy water of our streams and rivers after heavy rain or melting of snow. The breakdown into fine particles also takes place continually in undisturbed soil through the excretions of plant roots. Something arises like a sort of mineral milk. In this substance new minerals are formed which do not crystallise but remain in a state like curdled milk. This 'jelly' is the bearer of the above mentioned potential germinating forces of the mineral. This strange substance would 'like' to become crystalline, but remains in this state and is to be found in the soil together with the humus from plant residues. It has a strange attraction for, and unites with, the organic part of humus.

When the minerals take hold of plant substances in this way, a certain stability ensues which was not present in the individual substances while they were separate. Thus what comes from the plant 'ordered'.

There is still a third process. The 'certain stability' implies something incomplete, not yet finished. To bind this into a stable humus, the animal world must take part. The mineral-vegetable material is taken in and digested by certain lower animals. They are either true insect larvae or animals which remain throughout their lives at a larval (germinal) stage like the earthworm. In the digestive processes of these animals the mineral-vegetable substance is bound together and rounded off into a stable material by the co-operation of the germinating growth forces of the three kingdoms of nature, something quite new is formed which is the foundation of life for the whole Earth. Already here on Earth these forces work into the future.

If we observe these formative processes of the living soil and compare them with the picture of the macrocosmic process of which Rudolf Steiner spoke, it is not difficult to recognise the macrocosmic law which underlies the maintenance of life on the earth.

Thus even the mineral state appears as all-embracing embryonal force for the future just as much as the other kingdoms of nature. Human beings, however, are the 'salt of the Earth'; they are the guardian of all the germinating forces of the Earth. It is given to human beings to work hand in hand with the creators of the world for the future of the Earth. From a spiritual knowledge of the microcosmos and the macrocosmos they will learn to treat the Earth as do the creative forces of plant growth, which every spring enkindle anew the microcosmic salt process in root formation from which the new plant can grow. Then, from human husbanding of the Earth, a new Earth can arise in the future.

Table of Geological Formations

These tables, grouping the rocks into 12 sedimentary and 7 igneous formations, were made by Rudolf Steiner in 1890 when collaborating in the production of *Pierers Lexikon*, edited by J. Kürschner (reproduced from *Veröffentlichungen aus dem Literarischen Frühwerk*, Vol. IV, Book 19).

Additions by the author are in italics. The translators have omitted the names of a few minor local German series, unfamiliar to English-speaking readers.

A. Sedimentary

Era & Period	Rocks	Forms of Life
V Anthropozoic or Recent		
12. Quarternary	b) Alluvium– marine and freshwater a) Diluvium– glacial and periglacial deposits	3 Large mammals including mammoth, cave bear, primitive man
IV Caenozoic		
11. Neogene (*Pliocene Miocene brown coal, limestone, salt, gypsum*)	c) Freshwater deposits b) Brackish water deposits a) Mediterranian marine deposits	2 Large mammals including mastodon, dinotherium, aceratherium, apes, giant lizards, palm, fig, elm, birch

Era & Period	Rocks	Forms of Life
10. Palaeogene (*Oligocene &* *Eocene basalt, oil*)	b) Limestones, sandstones, clays and marls	1 Large mammals including palaeotherium, nummulites, fucoids
II Mesozoic		
9. Cretaceous *(oil)*	c) Sandstones, clays, limestones, Quader Sandstone, Cenomanian, Turonian and Senonian Chalk, Gosau formation	Last belemnites and ammonites
	b) Limestones, sandstones, marls, gault, red marls	First deciduous trees
	a) Limestones, sandstones, clays, marls	Sponges, foraminifera, spatangites, ammonites, belemnites, rudistids
8. Jurassic	d) Wälderton transitional beds	Giant saurians
	c) Malm or White Jura, Oxford	Reef corals, first bony fish *(birds)*
	b) Dogger or Brown Jura	Marsupials
	a) Lias or Black Jura	Belemnites, ammonites, sea saurians, cryptogams
7. Rhaetic. Dolomite Mountains	Dachstein limestone and dolomite, Kössener shales	Oldest mammal remains (Microlestes antiquus)
6. Triassic	c) Lettenkohle Keuper *(marls, sandstones and* *gypsum)*	Bipedal dinosaurs, *crocodiles*

Era & Period	Rocks	Forms of Life
	b) *(Muschelkalk)* dolomite, gypsum, salt	Sea lilies, ceratites, first crayfish
	a) Bunter sandstone, conglomerates, marls	Giant horsetails, conifers, first traces of birds

II Palaeozoic

5. Permian	b) White and gray beds, Kupferschiefer	First amphibians heterocercal tailed ganoid fish
	a) Rotliegend or totliegendes *(salt)*	Ferns, palms, conifers
4. Carboniferous	b) Productive coal measures *(oil)*	Terrestrial cryptogams, first conifers, spiders and insects
	a) Millstone grit *(culm)* Mountain Limestone *(oil)*	Last trilobites, productus
3. Devonian *(oil)*	c) Shelly limestone, cyprid slates, Old Red Sandstone	Armoured fish
	b) Eifel limestone	Brachiopods
	a) Graywackes	Primitive land plants, corals
2. Silurian	d) Upper silurian, limestones, slates	Seaweeds, corals, sea-lilies, brachiopods, first traces of fish
	c) Lower silurian, quartzites, shales *(oil)*	

Era & Period	Rocks	Forms of Life
	b) Ordovician, graywackes and slates	Trilobites, braciopods
	a) Cambrian, conglomerates, quartzites, shales	First unmistakable organic remains

I Pre-Cambrian

	d) Primal slates	Eozoon canadense
	c) Mica schists	in early limestone,
	b) Laurentian gneisses	if organic, is the
	a) Bojische gneisses	oldest fossil

B. Igneous

III Caenolithic	7	Younger trachytes and basalts
II Mesolithic	6	Younger porphyries
	5	Younger greenstones
	4	Younger granites
III Paleolithic	3	Older porphyries
	2	Older greenstones
	1	Older granites

Photo References

1. A little fold in gneiss. La Chaume, Sables d'Olonne, France. Arlette. Wikicommons

2. Rhyolite porphyry. ElfOwl245

3. Orbicular granite. Polished slab from Savitaipale, south eastern Finland. kristallin.de

4. Orbicular granite (orbicules 7 cm across). Caldera beach, Copiapo, Chili. Pierre J-P Bachelot, 2014

5. Mica rose (lithium mica) from pegmatite vein in granite. Schenkenzell, Black Forest, Germany. Walther Cloos

6. Muscovite, a potassium mica. Linópolis, Minas Gerais, Southeast Brazil. Rob Lavinsky

7. Fans of hornblende crystals with garnets. Hawley State Forest, Massachusetts. Colleen Sculleigh Osman

8. A cut surface of serpentine (unknown location). Mike Nelson

9. Asbestos (antigorite and picrolite) from Jeffrey Mine, Quebec, Canada. Michael P. Klimetz

10. Moss agate (thin cut slice)

11. Fine octahedrite internal structure of a Gibeon iron meteorite, found in Namaqualand, Namibia. Heritage Auctions

12. Pebbles with impressions from Myocene Torton Flysch. Rechtis, Kempten, South Germany. Walther Cloos

13. Basalt columns on the Isle of Staffa, Scotland. Iona Robertson, 2006

14. Basalt columns at Litlanesfoss, Iceland. Xfigpower on Wikimedia, 2011

Bibliography

Kolisko, Lilli, *Moon and Plant Growth,* Kolisko Archive Publications.
— Spirit in Matter. Stroud, 1948.
Laatsch, Willi, 'Untersuchungen über die Bildung und Anreicherung von Humusstoffen' in *Berichte über Landtechnik,* No. 4, 1948.
Pelikan, Wilhelm, *The Secrets of Metals,* Anthroposophic Press, USA 1973.
Pfeiffer, E. 'Die geologische Erdenentstehung im Lichte der Geisteswissenschaft,' in *Gäa-Sophia, Jahrbuch der naturwissenschaftlichen Sektion,* Vol. 1, Dornach 1926.
Steiner, Rudolf. Volume Nos refer to the Collected Works (CW), or to the German Gesamtausgabe (GA).
—, *Agriculture: Spiritual Foundations for the Renewal of Agriculture,* Biodynamic Farming and Gardening Association, USA 1993.
— *Anthroposophical Leading Thoughts* (CW 26) RSP 1973.
—, *Cosmic Memory* (CW 11) Steinerbooks, USA 1987.
—, *From Comets to Cocaine ... Answers to Questions* (CW 348) RSP 2002.
—, *From Crystals to Crocodiles ... Answers to Questions* (CW 347) RSP 2002.
—, *From Sunspots to Strawberries ... Answers to Questions* (CW 354) RSP 2002.
—, *Mensch und Welt, Das Wirken des Geistes in der Natur* (GA 351) Dornach 1999.
—, *Mystery Knowledge and Mystery Centres* (CW 232) RSP 2013.
—, *Occult Science: an Outline* (CW 13) RSP 2013 (also translated as *an Outline of Esoteric Science*).
—, *An Outline of Esoteric Science* (CW 13) AP 1997 (also translated as *Occult Science: an Outline*).
—, *Veröffentlichungen aus dem Literarischen Frühwerk,* Vol. IV, Dornach 1941.
—, *The World of the Senses and the World of the Spirit* (CW 134) RSP 2014.
Suess, Eduard, *The Face of the Earth,* 3 vols. Oxford 1904.
Vreede, Elisabeth, *Astronomy and Spiritual Science,* SB 2007.
Wachsmuth, Guenther, *Erde und Mensch, ihre Bildekräfte, Rhythmen und Lebensprozesse,* Constance 1952.
—, *Die Entwicklung der Erde, Kosmogonie und Erdgeschichte, ein organisches Werden,* Dornach 1950.

Index

Laatsch, Willi 45
labradorite 120
lamprophyre 53f
larvae 42f
lava 19f, 99, 103
— lakes 100
lead 71f, 130, 138f
Lemuria 157f
lepidolite 122
leucite 120
life, origin of 15, 23
lime 70
limestone 18, 20f, 29, 65–67, 119

magma 99
magnesium 40, 125
magnetic iron ore 128
manganese 133f
marble 66
margarite 122
marl 67f, 95
melaphyre, amigdaloidal 55f
mercury 130, 138f
mesenchyme 143
mesoderm 143
metals 144
meteoric 116
— dust 114
— iron 113, 115
— stone 113, 115f, 118
meteorite 112–17
meteoritic stone 113
meteors 112, 114, 117
mica 17f, 20, 38, 45, 55, 57f, 119, 122
mica roses 37
mica schist 17
mineral-plant 32f, 35–38, 40, 42– 45, 48f, 53
molybdenum 83, 137

morula 142
moss-agate 39
Muschelkalk 67–69
muscovite 122

nepheline 120
neptunists 14
nickel 83, 127
norite, orbicular 55
nosean 120
nuclear energy 152

oil 81–88
— bacteria 82, 84
— -bearing shales 85
— -digesting bacteria 83
Old Moon 27f, 32–34, 141
Old Saturn 27f, 141
Old Sun 27f, 141
oligoclase 120
olivine 116, 119, 122, 125
orthoclase 120f
ossification 142, 145
oyster 69

pahoehoe 104
paragonite 122
pegmatite 51–54
Pelé's hair 100
perthite feldspars 120
phlogopite 122
phonolite 102
phosphorite 71
phyllites 58
pitchblende 107
plagioclase 121
plant-animal 26, 33, 41–43, 47–50, 54, 56
Platonic Year 157f
plutonists 14
plutonium 152

Floris
Books

For news on all our **latest books,**
and to receive **exclusive discounts,**
join our mailing list at:

florisbooks.co.uk

Plus subscribers get a FREE book
with every online order!

Printed in the USA
CPSIA information can be obtained
at www.ICGtesting.com
JSHW012149311024
72768JS00003B/12